03-CTS-21UR

CONTRIBUTIONS OF HOUSEHOLD CHEMICALS TO SEWAGE AND THEIR RELEVANCE TO MUNICIPAL WASTEWATER SYSTEMS AND THE ENVIRONMENT

by:

Jörg E. Drewes

Eric Dickenson

Colorado School of Mines

Shane Snyder

Southern Nevada Water Authority

2009

WERF

WATEREUSE FOUNDATION

The Water Environment Research Foundation, a not-for-profit organization, funds and manages water quality research for its subscribers through a diverse public-private partnership between municipal utilities, corporations, academia, industry, and the federal government. WERF subscribers include municipal and regional water and wastewater utilities, industrial corporations, environmental engineering firms, and others that share a commitment to cost-effective water quality solutions. WERF is dedicated to advancing science and technology addressing water quality issues as they impact water resources, the atmosphere, the lands, and quality of life.

For more information, contact:
Water Environment Research Foundation
635 Slaters Lane, Suite 300
Alexandria, VA 22314-1177
Tel: (703) 684-2470   Fax: (703) 299-0742
www.werf.org   werf@werf.org

This report was co-published by the following organizations. For nonsubscriber sales information, contact:

IWA Publishing
Alliance House, 12 Caxton Street
London SW1H 0QS, United Kingdom
Tel: +44 (0) 20 7654 5500
Fax: +44 (0) 20 7654 5555
www.iwapublishing.com
publications@iwap.co.uk

WateReuse Foundation
1199 North Fairfax Street, Suite 410
Alexandria, VA 22314
Tel: (703) 548-0880
Fax: (703) 548-5085
www.WateReuse.org/Foundation
foundation@watereuse.org

Library of Congress Catalog Card Number: 2008939836

IWAP ISBN: 978-1-84339-529-4

Colorado School of Mines and Southern Nevada Water Authority

The research on which this report is based was developed, in part, by the United States Environmental Protection Agency (EPA) through Cooperative Agreement No. CR-831559-01with the Water Environment Research Foundation (WERF). However, the views expressed in this document are solely those of Colorado School of Mines and Southern Nevada Water Authority and neither EPA nor WERF endorses any products or commercial services mentioned in this publication. This report is a publication of WERF, not EPA. Funds awarded under the Cooperative Agreement cited above were not used for editorial services, reproduction, printing, or distribution.

ACKNOWLEDGMENTS

The research teams thanks the facilities that agreed to participate in this study. Their willingness to provide information and samples was of obvious importance to the success of this project. The team also would like to thank members of the project subcommittee (PSC) for their constructive critique and feedback. The individuals volunteering their time in this advisory role are listed below. Thanks also goes to Dr. Brad Price and his staff for analytical support provided throughout the study allowing us to broaden the range of compounds targeted in this study. Last, the team thanks the WERF program managers for their administrative assistance throughout the study.

Report Preparation

Principal Investigators:
Jörg E. Drewes, Ph.D.
Colorado School of Mines

Shane Snyder, Ph.D.
Southern Nevada Water Authority

Project Team:
Eric Dickenson, Ph.D.
Terry Jennings, M.S.
Tanja Rauch-Williams, Ph.D.
Colorado School of Mines

Brett Vanderford
Rebecca Trenholm
Dawn Lei, Ph.D.
Southern Nevada Water Authority

Project Subcommittee

Brad Price, Ph.D.
The Procter & Gamble Company

Rich Sedlak
The Soap and Detergent Association

Gregory Sales, Ph.D.
U.S. EPA National Risk Management Research Laboratory

Water Environment Research Foundation Staff

Director of Research:	Daniel M. Woltering, Ph.D.
Program Directors:	Christy Bixler, P.E.
	Margaret Stewart

ABSTRACT AND BENEFITS

Abstract:

Of the total number of consumer product chemicals the U.S. Environmental Protection Agency has identified, approximately 500 are considered high-production-volume (HPV) chemicals. This study investigated the occurrence and fate of HPV household chemicals in wastewater systems. The study was initiated with a comprehensive review on HPV organic chemicals in household commodities and their contributions to municipal wastewater treatment systems. The comprehensive review provided a basis to compile a database on HPV chemicals and organic compounds in household commodities that have the potential to affect wastewater processes and effluent qualities. The occurrence of select HPV target compounds during wastewater treatment was studied by collecting composite samples of raw sewage and final treated effluents at seven full-scale treatment plants employing different operational conditions. Of the 26 household chemicals targeted in this study, 20 compounds were consistently detected in raw influents of full-scale wastewater treatment plants. Chemicals that are primarily used in products applied outdoors were generally not present in raw influent samples. The majority of compounds present in personal care and cleaning products generally appeared in all influent samples with concentrations of 2-phenoxyethanol (a preservative with various uses) and menthol (a fragrance with various uses) consistently exhibiting the highest concentrations of all compounds. The ability of advanced wastewater treatment processes (i.e., membrane bioreactor, ozonation, advanced oxidation processes) to remove and destroy selected target compounds was studied through controlled lab- and pilot-scale studies. In general, biological treatment resulted in partial or complete removal (> 80%), indicating that biological treatment is a good treatment option for HPV household chemicals.

Benefits:

- Provides a comprehensive compilation of HPV organic chemicals in household products;
- Provides a validated framework how to identify household chemicals that potentially could contribute to sewage;
- Provides assurance that biological and chemical wastewater treatment processes are capable of achieving significant removal of household chemicals targeted; and
- Examines quantitative structural property relationship models regarding their ability to predict the fate of organic chemicals through specific treatment processes.

Keywords: Household chemicals, high-production-volume chemicals, wastewater treatment, fate and transport, microconstituents

TABLE OF CONTENTS

LIST OF TABLES

LIST OF FIGURES

LIST OF ACRONYMS

AOP	Advanced oxidation processes
APEC	Alkylphenol polyethoxycarboxylates
APEs	Alkylphenol polyethoxylates
AS	Activated sludge
BOD	Biochemical oxygen demand
BPA	Bisphenol A
CAS	Chemical Abstracts Service
CEPA	Canadian Environmental Protection Act
CERCLA	Comprehensive Environmental Response, Compensation and Liabilities Act
CERI	Chemical Evaluation and Research Institute (Japan)
CPS	Consumer Product Safety
CSM	Colorado School of Mines
CUS	Chemical Update System
DEHP	Di-(2-ethylhexyl) phthalate
D_p	Diffusion coefficient
DI	Deionized water
DMA	Dimethylamine
DOC	Dissolved organic carbon
DSL	Domestic Substances List
EDCs	Endocrine-disrupting chemicals
EfOM	Effluent organic matter
EI	Electron impact ionization
EII	Environmental Impact Initiative
ESIS	European Chemical Substances Information System
EU	European Union
EUCLID	European Uniform Chemical Information Data Base
F/M	Food-to-microorganism ratio
FDA	U.S. Food and Drug Administration
F&DA	Canadian Food and Drugs Act
FIFRA	Insecticide, Fungicide, and Rodenticide Act
GAC	Granular activated carbon
GC/MS-MS	Gas chromatography with tandem mass spectrometry
HMIS®	Hazardous Materials Identification System
HPA	Hazardous Products Act (Canada)
HPLC	High-performance liquid chromatography
HPV	High production volume
HRT	Hydraulic retention time
HSDB	Hazardous Substance Data Bank

IARC	International Agency for Research on Cancer
ICCA	International Council of Chemical Associations
INCI	International Nomenclature of Cosmetic Ingredients
IUCLID	International Uniform Chemical Information Database
IUR	Inventory Update Rule
K_{ow}	Octanol-water partitioning coefficients
LAS	Linear alkylbenzene sulfonate
LC/MS-MS	Liquid chromatography with tandem mass spectrometry
LOD	Limit of detection
MBR	Membrane bioreactor
MF	Microfiltration
MGD	Million gallons per day
MLSS	Mixed liquor suspended solids
MSDS	Material Safety Data Sheet
MW	Molecular weight
MWCO	Molecular weight cut-off
n.d.	Below detection limits
NF	Nanofiltration
ng/L	nanogram per liter
NIH	National Institutes of Health
NLM	National Library of Medicine
NOM	Natural organic matter
NP	Nonylphenol
NP1EO	Nonylphenol monoethoxylate
NP2EO	Nonylphenol diethoxylate
NPCA	National Paint and Coatings Association
NSNR	New Substances Notification Regulations
OECD	Organization for Economic Cooperation and Development
OP	Octylphenol
OSHA	Occupational Safety and Health Administration
PAC	Powdered activated carbon
pK_a	Negative log of the acid dissociation constant
ppb	Parts per billion (= micrograms per liter)
ppt	Parts per trillion (= nanograms per liter)
PSC	Project Subcommittee
PSP	Canadian Product Safety Programme
QA	Quality assurance
QC	Quality control
QSAR	Quantitative structure activity relationship
QSPR	Quantitative structure property relationship
RCRA	Resource Conservation and Recovery Act

REACH	Registration, Evaluation and Authorisation of Chemicals
RO	Reverse osmosis
SAT	Soil-aquifer treatment
SDWA	Safe Drinking Water Act
SIDS	Screening Information Data Set
SNWA	Southern Nevada Water Authority
SRD	Source Ranking Database
SRS	Substance Registry System
SRT	Solids retention time
TKN	Total Kjeldahl nitrogen
TOC	Total organic carbon
TRI	Toxic Release Inventory
TSCA	Toxic Substances Control Act
TSCA IUR	U.S. EPA's Toxic Substances Control Act Inventory Update Rule
TSS	Total suspended solids
UF	Ultrafiltration
UNEP	United Nations Environmental Program
U.S. EPA	U.S. Environmental Protection Agency
USHPV	The United States High Production Volume database
UV	Ultraviolet radiation
WHO	World Health Organization

EXECUTIVE SUMMARY

ES.1 Introduction

Households regularly use products containing inorganic and organic compounds that ultimately end up in municipal wastewater systems and, therefore, represent a potential threat to the environment. Of the total number of consumer product chemicals the U.S. Environmental Protection Agency has identified, approximately 500 are considered high-production-volume (HPV) chemicals. There is, however, a serious gap in our knowledge base regarding both the toxicity and fate of HPV commodity chemicals. An increasing number of reports of reproductive disorders in aquatic wildlife residing below wastewater outfalls have caused concern among scientists and the public. The lack of information both on the safety of HPV chemicals and organic compounds used in household commodities and the amount of hazardous substances being released into the environment during use and disposal is posing a major challenge to the water industry.

ES.2 Objectives

This study was tailored to address four primary objectives related to occurrence and fate of HPV household chemicals and chemicals in household commodities in wastewater systems:

1. To identify organic compounds from households and household commodities potentially present in municipal wastewater based on high-volume production, consumption pattern, and physicochemical properties.
2. To determine occurrence and fate of selected HPV target compounds through conventional and advanced wastewater treatment unit operations.
3. To compile a database on HPV chemicals and organic compounds in household commodities.
4. To identify and develop quantitative structure activity and property relationships (QSARs and QSPRs, respectively) that can be used to model the behavior of individual compounds through treatment processes.

ES.3 Research Approach

The study was initiated with a comprehensive review of HPV organic chemicals and organic compounds in household commodities and their contributions to municipal wastewater treatment systems. This survey provided the basis to select a short list of potential HPV organic compounds that were targeted during this study. The comprehensive review presented the basis to compile a database on HPV chemicals and organic compounds in household commodities that have the potential to affect wastewater processes and effluent qualities. The occurrence of select HPV target compounds during wastewater treatment was studied by collecting composite samples of raw sewage and final treated effluents at seven full-scale treatment plants employing different operational conditions. Additional efforts were directed to assess the removal efficiency of selected HPV chemicals by conventional unit operations (e.g., activated-sludge, disinfection). The ability of advanced wastewater treatment processes (i.e., membrane bioreactors, ozonation, advanced oxidation processes) to remove and destroy selected target compounds was studied in controlled lab- and pilot-scale studies.

After a comprehensive literature review on existing QSPR models that aid or could aid in predicting the removal of certain compounds by individual unit operations relevant to wastewater treatment, gaps in existing QSPR efforts were identified regarding both not represented removal mechanisms (e.g., chlorination) and specific treatment processes. Efforts were directed to derive descriptive parameters through controlled activated-carbon laboratory studies to improve QSPR modeling efforts. The final outcome of these efforts identified the most viable QSPR components, which can be connected as modules – as part of a toolbox – to predict the removal of individual compounds through wastewater treatment trains.

ES.4 Selection of HPV Household Chemicals

A comprehensive list of household chemicals was initially derived from a database offered by the National Institutes of Health (NIH) National Library of Medicine (NLM). This was the most comprehensive database identified covering different household commodities. From this database, approximately 720 compounds were identified that fall within eight commodities: 1) auto products, 2) inside the home, 3) pesticides, 4) home maintenance, 5) personal care/use, 6) pet care, 7) arts and crafts, and 8) landscape/yard. Inorganic and organic compounds were considered from this database even though only organic compounds are targeted in this study. In general, most drugs and food additives were not considered.

The selection of compounds from the comprehensive list of approximately 720 compounds identified for further study and analysis followed a two-tiered ranking approach based on production volumes, environmental relevance, and feasibility for analytical quantification. The selection of Tier 1 chemicals included only single compounds (vs. compound mixtures) and those listed in the U.S. EPA's Toxic Substances Control Act Inventory Update Rule (TSCA IUR) database with yearly production volumes greater than 1 million pounds per year and which are likely present in domestic wastewater due to their physicochemical properties and reported environmental fate. The short list of 11 Tier 1 chemicals targeted in this study for monitoring is presented in Table ES-1. Tier 2 chemicals included compounds with production volumes less than one million pounds per year, or for which no specific production volume was identified, even though these compounds were identified in at least one of the three volume production databases considered in this study. Tier 2 chemicals are also frequently used in household commodities and are also likely to be present in domestic wastewater due to their physicochemical properties and reported environmental fate. The short list of 13 Tier 2 chemicals targeted in this study appears in Table ES-1. Analytical methods were developed by the research team for the Tier 1 and 2 compounds using liquid chromatography with tandem mass spectroscopy (LC/MS-MS) and gas chromatography with tandem mass spectroscopy (GC/MS-MS). Linear alkylbenzene sulfonate (LAS) and triclocarban were added to the target compound list because LAS can function as a model for complex mixtures and triclocarban as a model for emerging contaminants for which only a limited amount of data was available at the time of study initiation. To study fate and transport of household chemicals covering a wide range of physicochemical properties during controlled laboratory-scale studies, the team also included several previously studied pharmaceutical residues to the list of compounds of interest.

Table ES-1. Target Compounds Selected in This Study.

Compound	Synonyms	Application
Tier 1 Compounds		
2,6-Di-t-butyl-p-cresol (BHT)	2,6-Di-t-butyl-p-cresol; BHT; 2,6-Di-tert-butyl-p-cresol; Butylated hydroxytoluene; Butylhydroxytoluene; Di-t-butylcresol; Phenol, 2,6-bis(1,1-dimethylethyl)-4-methyl-; p-Cresol, 2,6-di-tert-butyl-	skin care, hobby supplies
Dibutyl phthalate	1,2-Benzenedicarboxylic acid, dibutyl ester; Butyl phthalate; DBP; Di-n-butyl phthalate; Dibutyl 1,2-benzenedicarboxylate; Dibutyl phthalate; Phthalic acid, dibutyl ester	plasticizer, finish, nail care
Atrazine	2-Chloro-4-(2-propylamino)-6-ethylamino-s-triazine; 2-Chloro-4-ethylamino-6-isopropylamino-s-triazine; 2-Chloro-4-ethylamino-6-isopropylamino-1,3,5-triazine; 2-Chloro-4-ethylamineisopropylamine-s-triazine; 2-Ethylamino-4-isopropylamino-6-chloro-s-triazine	Herbicide
4,4-Isopropylidenediphenol (Bisphenol A)	4,4'-(1-Methylethylidene)bisphenol; 4,4'-Bisphenol A; 4,4'-Isopropylidenediphenol; Bisphenol A; Diphenylolpropane; Phenol, 4,4'-(1-methylethylidene)bis-; Phenol, 4,4'-isopropylidenedi-; 4,4' Isopropylidinediphenol	plasticizer, epoxy, glue
Benzophenone	Benzophenone; Benzoylbenzene; Diphenyl ketone; Diphenylmethanone; Methanone, diphenyl-	car wash, facial products
Benzophenone-3 (Oxybenzone)	(2-Hydroxy-4-methoxyphenyl)phenylmethanone; 2-Hydroxy-4-methoxybenzophenone; Benzophenone-3; Methanone, (2-hydroxy-4-methoxyphenyl)phenyl-; Oxybenzone; Benzophenone, 2-hydroxy-4-methoxy-	skin care, insect repellent
Triclosan	Triclosan; Phenol, 5-chloro-2-(2,4-dichlorophenoxy)-; Ether, 2'-hydroxy-2,4,4'-trichlorodiphenyl; 5-Chloro-2-(2,4-dichlorophenoxy)phenol; 2,4,4'-Trichloro-2'-hydroxy diphenyl ether	antibacterial, soap, lotion
Vanillin	p-Hydroxy-m-methoxybenzaldehyde; 4-Hydroxy-m-anisaldehyde; 4-Hydroxy-3-methoxybenzaldehyde; Benzaldehyde, 4-hydroxy-3-methoxy-; Vanillin [USAN]	fragrance, cosmetics, various
o-Phenylphenol	(1,1'-Biphenyl)-2-ol; 1-Hydroxy-2-phenylbenzene; 2-Phenyl phenol; Biphenyl-2-ol; 2-Phenylphenol; Biphenyl, 2-hydroxy-; o-Phenylphenol	germicide, fungicide
2-Phenoxyethanol	Phenoxyethanol; 2-Phenoxyethanol; Ethanol, 2-phenoxy-; Ethylene glycol monophenyl ether; Phenoxytol	preservatives, cosmetics, fragrance, various
Hexabromocyclododecane	1,2,5,6,9,10-Hexabromocyclododecane; Hexabromocyclododecane; Cyclododecane, 1,2,5,6,9,10-hexabromo-	flame retardant
Tier 2 Compounds		
Simazine	1,3,5-Triazine-2,4-diamine, 6-chloro-N,N'-diethyl-; s-Triazine, 2-chloro-4,6-bis(ethylamino)-; Simazine	herbicide, anti-algae products
N,N-Diethyl-meta toluamide (DEET)	N,N-Diethyl-meta-toluamide; DEET; Benzamide, N,N-diethyl-3-methyl-; N,N-Diethyl-m-toluamide; m-Toluamide, N,N-diethyl-	insect repellent, industrial
Hydrocortisone	11-beta-Hydrocortisone; 11beta,17,21-Trihydroxypregn-4-ene-3,20-dione; 17-Hydroxycorticosterone; 17alpha-Hydroxycorticosterone; 4-Pregnene-11beta,17alpha,21-triol-3,20-dione; Cortisol; Hydrocortisone; Hydroxycortisone	Anti-itch cream
Butylated hydroxyanisole (BHA)	BHA; tert-Butyl-4-methoxyphenol; 3-tert-Butyl-4-hydroxyanisole; Anisole, butylated ydroxyl-; Phenol, (1,1-dimethylethyl)-4-methoxy-	Antioxidant, various
3-Indolebutyric acid	4-Indol-3-ylbutyric acid [BSI:ISO]; Indole-3-butyric acid; Butyric acid, 4-(indolyl)-; Indole butyric acid; Indole-3-butanoic acid	fertilizer, Miracle-Gro™ fertilizer
Camphor	2-Bornanone; 1,7,7-Trimethylbicyclo(2.2.1)heptan-2-one; Bicyclo(2.2.1)heptan-2-one, 1,7,7-trimethyl-; Camphor; Spirit of camphor; Gum camphor; (+-)-Camphor; DL-Bornan-2-one	fragrance, various
Menthol	Menthol; 5-Methyl-2-(1-methylethyl)cyclohexanol; 2-Isopropyl-5-methylcyclohexanol; Menthyl alcohol; Cyclohexanol, 5-methyl-2-(1-methylethyl)-; DL-Menthol	fragrance, various
2-Methylresorcinol	1,3-Benzenediol, 2-methyl-; Resorcinol, 2-methyl-	Hair color, cosmetics
Isobutylparaben	Benzoic acid, p-hydroxy-, isobutyl ester; Benzoic acid, 4-hydroxy-, 2-methylpropyl ester (9CI); Isobutyl 4-hydroxybenzoate; 2-Methylpropyl 4-hydroxybenzoate; Isobutylparaben; Isobutyl parahydroxybenzoate; 4-Hydroxybenzoic acid, 2-methylpropyl ester; Isobu	preservative, cosmetics, various
Acriflavine	Xanthacridinum; Acridinium, 3,6-diamino-10-methyl-, chloride, mixt. With 3,6- acridinediamine	Pesticide
Trifluralin	2,6-Dinitro-N,N-di-n-propyl-alpha,alpha,alpha-trifluoro-p-toluidine;	herbicide

Compound	Synonyms	Application
	Trifluralin; Benzenamine, 2,6-dinitro-N,N-dipropyl-4-(trifluoromethyl)-; N,N-Dipropyl-2,6-dinitro-4-trifluoromethylaniline; 4-(Trifluoromethyl)-2,6-dinitro-N,N-dipropylaniline; 2,6-Dinit	
2,3,4,5 Bis(2-butylene) tetrahydro 2-furaldehyde (MGK-11)	Dibutylene tetrafurfural; 2,3:4,5-Di(2-butenyl)tetrahydrofurfural; 1,4,4°,5°,6,9,9°,9b-Octahydrodibenzofuran-4°-carbaldehyde	insect repellent
Propylparaben	4-Hydroxybenzoic acid, propyl ester; Benzoic acid, 4-hydroxy-, propyl ester; Benzoic acid, p-hydroxy-, propyl ester; Propyl 4-hydroxybenzoate; Propyl parahydroxybenzoate; Propyl p-hydroxybenzoate; Propyl paraben	preservative, various
Model Compounds		
Triclocarban	1-(3',4'-dichlorophenyl)-3-(4'-chlorophenyl)urea; 3,4,4'-trichlorodiphenylurea; 3,4,4'-Trichlorocarbanilide; cusiter; cutisan; genoface; N-(3,4-dichlorophenyl)-N'-(4-chlorophenyl)urea; N-(4-chlorophenyl)-N'-(3,4-dichlorophenyl)urea; procutene; Solubacter; TCC; Trichlorocarbanilide; triclocarban	antibacterial, soap, deodorant
Linear alkylbenzene sulfonate (LAS)	n.a.	surfactant, various
Indicator Compounds		
Primidone		Antiepileptic drug
Phenacetine		Anti-inflammatory drug
Carbamazepine		Antiepileptic drug
2-Naphthol		Industrial chemical
Fenofibrate		Blood lipid regulator
Gemfibrozil		Blood lipid regulator
Propyphenazone		Anti-inflammatory drug
Sulfamethoxazole		Antibiotic drug
Ibuprofen		Anti-inflammatory drug
Naproxen		Anti-inflammatory drug
Diclofenac		Anti-inflammatory drug

ES.5 Full-Scale Facilities

Field investigations for this study were carried out at six participating facilities located in three different states within the western continental United States. This approach allowed an investigation of both regional variations of effluent quality and the removal effectiveness of individual processes comprising an overall process train. During the initial phase of the study, the research team conducted a survey at all participating utilities to determine unit processes and operations employed, process parameters, water qualities, and service area characteristics for each individual wastewater treatment plant participating in this study. The survey represented the basis for selecting field sites, unit operations, and sampling logistics for full-scale sampling. The participating utilities assisted with sampling logistics and provided operational information and data for the time of sampling.

During field-scale occurrence monitoring, raw influent and final effluent samples were collected over a 24-hour period usually commencing Tuesday morning until the following morning. Samples were collected using existing dedicated facility autosamplers used for quality and compliance purposes. Samples were collected directly into a single, large 10-L glass container housed in a refrigerated cabinet. All samples were glass-fiber filtered on site at the end of the collection period and collected into 1-L ultra-clean, pre-silanized, amber glass bottles containing sodium azide for preservation and ascorbic acid solution to quench any residual chlorine. Samples were then placed in a specialized hard-shell shipping case with ice packs and shipped, via overnight courier, to the analytical laboratory. All samples were extracted within 14 days of collection. Operational and water quality parameters measured at the time of sampling were compiled by the project team at the Colorado School of Mines (CSM).

ES.6 Occurrence of HPV Household Chemicals in Raw Sewage and Wastewater Effluents

Of the 26 household chemicals targeted in this study, 20 compounds were consistently detected in raw influents of full-scale wastewater treatment plants. Chemicals that are primarily used in products applied outdoors, such as pesticides (e.g., simazine, atrazine, trifluralin, and acriflavine), were generally not present in raw influent samples. The majority of compounds present in personal care and cleaning products generally appeared in all influent samples with concentrations of 2-phenoxyethanol (a preservative with various uses) and menthol (a fragrance with various uses) consistently exhibiting the highest concentrations of all compounds.

Occurrence in glass-fiber filtered raw sewage and treated effluent among the six facilities showed very similar concentration ranges, supporting the hypothesis that the studied HPV compounds are contributed by general consumers/households, rather than certain industries. The concept employed in this study for selecting HPV chemicals was confirmed, where, in general, concentrations of Tier 1 compounds in raw sewage were in the microgram-per-liter range, whereas Tier 2 compounds were present at several hundred nanograms-per-liter. In treated wastewater effluents, the Tier 1 compounds dibutyl phthalate, benzophenone, 2-phenoxy-ethanol, vanillin, and triclosan exhibited average effluent concentrations exceeding 200 ng/L with dibutyl phthalate exhibiting the highest average concentration of approximately 600 ng/L. Only three out of 13 Tier 2 compounds, 3-indolebutyric acid, DEET, and menthol, exhibited concentrations exceeding 100 ng/L in secondary treated effluents. Most of the compounds occurred at concentrations of less than 200 ng/L, which is comparable to the occurrence of pharmaceutical residues. The average concentration of LAS in raw sewage almost approached 6 mg/L and concentration was reduced in secondary treated effluents to approximately 4,000 ng/L, representing the highest observed concentrations of all compounds targeted in this study.

Daily mass loading variations of all Tier 1 compounds suggested a rather constant discharge of household chemicals to sewage with discharge directly coupled to water consumption and thus linked to consumer discharge. Daily mass loadings of Tier 2 compounds were significantly lower than Tier 1 compounds but also followed the general diurnal trend. Higher weekend loading values for some compounds (i.e., dibutyl phthalate, oxybenzone, triclocarban, camphor, and DEET) may further indicate that many of these compounds have significant usage in household products. Conversely, lower weekend loading values observed for benzophenone and o-phenylphenol may indicate that those compounds also have significant uses other than in households (i.e., industrial). Daily LAS mass loading exhibited a pattern following the common diurnal flow regime. This finding is expected because LAS is widely used as a surfactant in many cleaning products and its use seems to be closely coupled with human activities and water consumption. Weekly sampling, however, revealed a significant higher loading of LAS during the weekend, which might be a consequence of more cleaning/laundry activities during the weekend.

Occurrence in the aqueous phase of sewage seems to be not only dependent upon production volume but also dependent upon compound properties. Of the three compounds produced between 10 and 50 million pounds per year (BHT, dibutyl phthalate, and 2-

phenoxyethanol), 2-phenoxyethanol is the most water soluble (log K_{ow} = 1.16) exhibiting the highest mass loading in raw sewage. BHT is the most hydrophobic compound (log K_{ow} = 5.1) and exhibited the lowest aqueous mass loading. A similar trend was observed for compounds produced between 1 and 10 million lb/yr, with vanillin being the most hydrophilic (log K_{ow} = 1.21), exhibiting the highest mass loading and triclocarban, being the most hydrophobic (log K_{ow} = 4.9), exhibiting the lowest aqueous phase mass loading. For compounds produced in smaller volume bins, no clear relationship between production volume and hydrophobicity was observed. Other HPV chemicals not addressed in this study that have similar properties to compounds detected in raw wastewaters in this study are also likely present in the raw water. An in-depth analysis (e.g., QSPR) of the properties of compounds examined in this study will help in predicting how many other HPV chemicals will be present and perhaps at what concentrations, in raw wastewaters.

ES.7 Fate of HPV Household Chemicals during Advanced Unit Processes

The removal or transformation of HPV chemicals by biological treatment (activated-sludge and membrane bioreactor [MBR]), disinfection (ultraviolet [UV] light, chloramines, chlorine), and advanced oxidation (ozone and ozone/hydrogen peroxide) processes was evaluated. In general MBR and activated-sludge treatments removed the Tier 1 and 2 compounds similarly. All Tier 1 and 2 compounds that were detected were partially or completely removed (> 80%), indicating that biological treatment is a good treatment option for these compounds. Of the conventional disinfection processes, chlorination represents a process for partial transformation of Tier 1 and 2 compounds and a second barrier. However, it is noted that likely halogenated by-products were formed during this process. Chloramine and UV disinfection processes are not viable removal barriers. Ozonation and ozone/hydrogen peroxide treatments performed similarly and are additional barrier options where higher effluent quality is desired. However, they do not represent a 100% barrier for all compounds (e.g., vanillin).

ES.8 QSPR Models to Predict HPV Compound Behavior

QSPRs are viable empirical tools that can be coupled with mass balance models to quickly assess the fate and treatment of new and existing chemicals of concern within wastewater treatment systems. The removal of an organic chemical is dependent upon the compound's structure. There is a broad range of well-defined molecular descriptors (e.g., 2D- and 3D-based descriptors) that are easily calculable and potentially relevant to linking the structural features of a compound to its ability to be removed by a particular treatment mechanism. Several existing QSPRs that are relevant to significant wastewater treatment removal routes, reactivity, phase partitioning, and physical separation, have been identified as viable tools that can be applied to the prediction of the removal of individual compounds through wastewater treatment trains. However, there is a need to evaluate QSPRs for their applicability to a wide range of organic compounds and realistic environmental conditions. Also, QSPRs need to be evaluated in regards to which QSPRs best describe chemically and/or physically an organic compound's potential to be removed by a particular removal mechanism. For some removal mechanisms there is a lack of uniformly measured treatment data (e.g., biodegradation rate constants). Therefore, there is a need to generate fate parameter data, particularly biodegradation and chlorination removal rates, for the evaluation of existing and development of new applicable QSPR models. Finally, QSPRs need to be incorporated into appropriate mass balance models, and these mass balance models (e.g., fugacity models) need to be validated with full-scale treatment data.

CHAPTER 1.0

INTRODUCTION

1.1 Introduction

Households regularly use products containing organic compounds that ultimately end up in municipal wastewater systems and, therefore, represent a potential exposure of these compounds to the environment. The average household typically uses and stores more than 60 hazardous products, including household cleaners, automotive products, paints, solvents, and pesticides (*Seattle Daily Journal of Commerce,* 1997). Of the total number of consumer product chemicals the U.S. Environmental Protection Agency (U.S. EPA) has identified, approximately 500 are considered high-production-volume (HPV) chemicals. There is, however, a serious gap in our knowledge base regarding both the toxicity and fate of HPV commodity chemicals. An increasing number of reports of reproductive disorders in aquatic wildlife residing below wastewater outfalls have caused concern among scientist, utilities, and the general public. The public also frequently perceives that chemicals are responsible for causing numerous diseases. Current discussions have focused on bisphenol A, phthalates, and brominated flame retardants, to name just a few. The lack of information both on the safety of HPV chemicals and organic compounds used in household commodities and the amount of hazardous substances being released into the environment during use and disposal is posing a major challenge to the water industry. This study was designed to close this knowledge gap.

1.2 Objectives

This study was designed to address four primary objectives related to occurrence and fate of household chemicals and chemicals in household commodities in wastewater systems. These four objectives are listed below with a brief description of the research approach selected by the project team. The objectives were:

1. To identify organic compounds from households and household commodities potentially present in municipal wastewater based on high-volume production, consumption pattern, and physicochemical properties.

This objective was addressed by conducting a comprehensive review on the current status of knowledge regarding occurrence of HPV organic chemicals and organic compounds in household commodities. This survey provided the basis to select a short list of potential HPV organic compounds from households and household commodities that were targeted during this study.

2. To determine occurrence and fate of selected HPV target compounds through conventional and advanced wastewater treatment unit operations.

Fate and transport of select HPV target compounds during wastewater treatment was studied by collecting composite samples of raw sewage and treated effluents at various full-scale treatment facilities employing different operational conditions. In addition, the removal efficiency of selected target compounds by individual conventional unit operations was assessed at full-scale facilities. The ability of advanced wastewater treatment processes to remove and destroy selected target compounds was studied through controlled bench- and pilot-scale studies (i.e., advanced oxidation, activated carbon).

3. To compile a database on HPV chemicals and organic compounds in household commodities.

The project team compiled a database on HPV chemicals and organic compounds in household commodities that have the potential to affect wastewater processes and effluent quality.

4. To develop quantitative structure activity and property relationships (QSAR/QSPR) which can be used to model the behavior of individual compounds through treatment processes.

The team conducted a comprehensive literature review on existing QSAR and QSPR models designed to predict removal of certain compounds by individual unit operations relevant to wastewater treatment. This review identified gaps in existing modeling efforts for removal mechanisms and specific treatment processes. Efforts were directed to derive descriptive parameters through controlled laboratory studies to improve QSPR modeling efforts. In parallel, the team identified and evaluated commercially available QSPRs. The final outcome of these efforts identified the most viable QSPR components, which can be connected as modules – as part of a toolbox – to aid in predicting the removal of individual compounds through wastewater treatment trains.

1.3 Background and Significance

1.3.1 Household Chemicals

1.3.1.1 Administration of Household Chemicals in the U.S.

In the United States, legislation for modern chemicals passed in the late 1970s instituting rigorous data and review requirements for substances placed on the market after 1980. Those chemicals on the market prior to that time (representing more than 99% by volume) were "grandfathered" with few regulatory requirements.

The U.S. EPA is the federal agency responsible for regulation of most chemicals that can enter the environment. Among the Acts that U.S. EPA administers are the Insecticide, Fungicide, and Rodenticide Act (FIFRA), the Toxic Substances Control Act (TSCA), the Resource Conservation and Recovery Act (RCRA), the Safe Drinking Water Act (SDWA), and the Comprehensive Environmental Response, Compensation and Liabilities Act (CERCLA) (Superfund Act). The U.S. Food and Drug Administration (FDA) is responsible for the safety evaluation of drugs, cosmetics, food additives, and medical devices. U.S. EPA does not explicitly track production rates of household chemicals. However, several databases are available that track the production, management, release, and use of certain chemicals.

U.S. EPA promulgated the Inventory Update Rule (IUR) in 1986 for the partial updating of the TSCA Chemical Substances Inventory database. The rule requires manufacturers and importers of certain chemical substances included on the TSCA database to report current data on the production volume of these substances. Starting in 1986, under the IUR, chemical manufacturers and importers provide U.S. EPA every four years with the relevant information pertaining to their most recent fiscal year. Manufacturers and importers of chemicals already on the TSCA database that are being produced at one plant site or imported at production volume levels of 10,000 or more pounds must report, unless either their chemical is excluded from the IUR or they fall under the small business exemption (EPA, 2003). The database can be searched by chemical name or CAS number and gives aggregated production rates for the four-year time ranges between 1986 and 2002 (www.epa.gov/opptintr/iur).

The U.S. EPA is developing the Source Ranking Database (S.R.D.), which identifies the sources of consumer product chemicals (U.S. EPA, 1998). Among 1,303 chemicals in the S.R.D., 491 are HPV chemicals. Chemicals considered to be HPV in the U.S. are those that are manufactured in or imported into the United States in amounts equal to or greater than 1 million pounds per year.

The Hazardous Materials Identification System (HMIS®) is a registered trade name of the National Paint and Coatings Association (NPCA) and is used by all manufacturers to comply with the requirements of Occupational Safety and Health Administration's (OSHA) Hazard Communication Standard. HMIS® uses colored bars, numbers, and symbols to convey the severity of hazards of chemicals. Only numbers are used in this database to convey the severity of the hazards of chemical ingredients of each brand when used in the workplace. These ratings were determined for each brand by its manufacturer and documented in the Material Safety Data Sheet (MSDS) published by the manufacturer (NIH, 2004).

The National Institutes of Health (NIH, 2004) offers an online household product database to inform consumers about ingredients and health effects of consumer products in the area of auto, cleaning, pesticides, landscape/yard, personal care, home maintenance, arts and crafts, and pet care. This service is based on the MSDSs, which are provided by the manufacturers of the consumer products and include a list of the ingredients and their CAS numbers. Information on proper disposal is included, but no information on environmental fate or produced amounts of specific chemicals is given. Revealing product information is voluntary. Therefore, this database is limited by the availability and quality of the data provided by the distributor/manufacturer. This list was initially compiled in 1995 and has since been updated by the contractor. Products included in the database are selected by market share and shelf presence in retail stores. This database can be screened for information on what chemicals appear in common household products. The database also offers a list of products that contain a certain chemical and the approximate concentrations. This database does not include information on drugs or food additives. Information on drugs can be retrieved from NIH (2004) and the U.S. National Library of Medicine (www.nlm.nih.gov/medlineplus/druginformation.html) and the U.S. FDA (2004b) (www.fda.gov/cder/drug/default.htm). With regard to medical drugs, the pharmaceutical industry offers information on the top-selling drugs in the United States. (https://ecom.nhin.com/nhin/servlet/DrugSearchEntry?CHAIN_ID=119080). In 2007, the list contains currently approximately 180 pharmaceutical drugs, their common uses, as well as medical information on the product. There is no information included on the major active components of the drugs. This information can be retrieved from the *Orange Book,* published electronically by the FDA (FDA, 2004a) (www.fda.gov/cder/ob/default.htm). The *Orange Book* Query enables searching of the approved drug list by active ingredient. The data are updated concurrently. Information on food additives is provided by the FDA in Title 21 Food and Drugs (www.access.gpo.gov/nara/cfr/waisidx_01/21cfrv3_01.html).

In U.S. EPA's Substance Registry System (SRS), chemicals can usually be searched by name or CAS number to determine how they are represented in U.S. EPA regulations and information systems (www.epa.gov/srs). The database provides a list of regulations applicable to a chemical and other regulatory information, such as reasons for the regulation, effective date, and withdrawn data, and literature references (U.S. EPA, 2004a). References to numerous other information databases are given, including:

- ECOTOX is a comprehensive database providing adverse effects of single chemical stressors to ecologically relevant aquatic and terrestrial species.
- CUS (Chemical Update System) contains information that must be reported by chemical manufacturers and importers every 4 years as a partial update to the chemical inventory, which is maintained under the Toxic Substances Control Act (TSCA). CUS94 contains information about all chemicals for which the identity has not been claimed Confidential Business Information (CBI) in the 1994 Chemical Update System, including chemicals for which both CBI and non-CBI production volume aggregates are available.
- USHPV (The United States High Production Volume database) provides information about chemicals that are included in the High Production Volume Challenge Program.
- HSDB (Hazardous Substance Data Bank) is a factual, non-bibliographic data bank focusing upon the toxicology of potentially hazardous chemicals including emergency handling procedures, environmental fate, human exposure, industrial hygiene, and regulatory requirements. HSDB is being built, maintained, reviewed, and updated by the National Library of Medicine (NLM).

1.3.1.2 Production Volume and Toxicity

The U.S. EPA estimates that until now approximately 70,000 chemicals have been produced in the United States since the mid-1970s (U.S. EPA, 2002). Further, it is estimated that there are more than 15,000 HPV chemicals manufactured in or imported into the United States in amounts equal or greater than 1 million pounds per year. Within these 15,000 compounds, 3,000 are organic chemicals (excluding polymers) (U.S. EPA, 2002). U.S. EPA developed the HPV Challenge Program to complete baseline testing on HPV chemicals by the year 2004. Through this program, U.S. EPA is challenging the chemical industry to undertake testing on HPV chemicals voluntarily. However, U.S. EPA will eventually mandate testing of all HPV chemicals by law under the testing authority of Section 4 of the Toxic Substance Control Act (TSCA).

For about 75% of the 15,000 HPV chemicals in commercial use, government authorities both in the United States and Europe have not conducted even basic toxicity testing. More than 90% of these HPV chemicals have not been tested for health effects on children who are often more vulnerable to impacts caused by toxic substances due to their immature and rapidly growing physiologic systems and greater exposure risks (Wargo and Wargo, 2002). Approximately 43% of the 3,000 organic HPV chemicals have never been tested for their toxicity or environmental impact considering six hazardous endpoints set by the United Nations Environmental Program (UNEP) in the Screening Information Data Set (SIDS) (Landrigan, 2000; EPA, 2002).

In 1987, U.S. EPA initiated the Toxic Release Inventory (TRI) to provide information to the public on releases and other waste management information for more than 600 chemicals with a high exposure potential for certain industry sectors (www.epa.gov/tri/tridata/tri02/data/index.htm) (U.S. EPA, 2004b). Of this group, 91 were high-release HPV TRI chemicals, out of which 74% had information available for all six basic SIDS tests (U.S. EPA, 1998). This database gives information by state or region on chemicals and their specific emission pathways and amounts.

The Chemical Economics Handbook is a comprehensive source of information on consumption patterns, demand projections, environmental issues, and production rates of about 300 industrial chemicals and chemical groups published by the Stanford Research Institute (2004). The hard copy comprises 38 binders with monthly updates for subscribers. Reports are available individually online for a fee (http://ceh.sric.sri.com).

1.3.1.3 Environmental Fate

Finding aggregated and comparable information on environmental fate of certain chemicals is challenging. One reason for this is that research studies often vary widely in their design and findings (e.g., with regard to biodegradation, adsorption, rates of removal) and are thus difficult to generalize. The Syracuse Research Corporation, a nonprofit, independent, research organization provides different comprehensive databases on the environmental fate of chemicals that can be searched online via CAS number (Syracuse Research Corporation, 2004) (www.syrres.com/esc/efdb.htm).

- DATALOG is a bibliographic file containing 18 types of environmental fate data (e.g., adsorption, bioconcentration, biodegradation, dissociation constants, ecosystem, effluent concentrations, evaporation from water, Henry constant, partition coefficients, photo-oxidation, and water solubility).
- BIOLOG provides sources of microbial toxicity and biodegradation data and gives references for specific environments (e.g., anaerobic or aerobic conditions, soil, sewage, freshwater, sediments, and reaction pathways).
- CHEMFATE is a data value file containing 25 categories of environmental fate and physicochemical property information on commercially important chemical compounds. Among these, information is available on soil adsorption constants, soil column transport, degradation in natural systems, water monitoring, soil monitoring, and field sites.
- BIODEG contains experimental values relating to biodegradation subjects if available for a specific compound.

To a certain extent, conclusions on the environmental fate can be derived from measurable chemical and physical properties, such as the octanol-water partitioning coefficient (log K_{ow}), Henry's constant (H), water solubility (S), the acid dissociation constant (pK_a), and the chemical structure. There are several public-domain databases available providing this information such as ScienceFinder or Chemfinder (http://chemfinder.cambridgesoft.com).

The biodegradation potential of a certain chemical in the environment can be estimated to a certain extent by its chemical structure. The software tool BIOWIN (Syracuse) calculates the probability that a chemical under aerobic conditions with mixed cultures of microorganisms will biodegrade rapidly or slowly. CATABOL (LMC Bourgas, Bulgaria) is a mechanistic modeling approach for quantitative assessment of biodegradability in biodegradation pathways of chemicals. The system generates the most plausible biodegradation products and provides quantitative assessment for their solubility and toxic endpoints. The software model EPI Suite (U.S. EPA, 2000) incorporates a number of models to estimate environmental fate and transport of chemicals (www.epa.gov/opptintr/exposure/docs/episuitedl.htm):

- BCFWIN: estimates bioconcentration factor (BCF)
- BIOWIN: estimates biodegradation probability (see description above)

- University of Minnesota biodegradation/biotransformation database (http://umbbd.msi.umn.edu/)
- ECOSAR: estimates aquatic toxicity (LD_{50}, LC_{50})
- HENRYWIN: estimates Henry's law constant
- HYDROWIN: estimates aqueous hydrolysis rates (acid-, base-catalyzed)
- KOWWIN: estimates octanol-water partition coefficient
- MPBPWIN: estimates melting point, boiling point, and vapor pressure
- PCKOCWIN: estimates soil sorption coefficient (K_{oc})
- WSKOWWIN: estimates water solubility (from log octanol-water partition coefficient)
- WATERNT: estimates water solubility (using atom-fragment methodology).

1.3.1.4 Administration of Household Chemicals in Canada

Within the regulatory framework of Canada, the Hazardous Products Act (HPA) is important for the regulation of hazardous consumer products. The purpose of the act is to protect the health and safety of consumers by prohibiting or regulating the sale, advertising, or import of hazardous or potentially hazardous consumer products. The act regulates consumer chemical products classified as very toxic, very corrosive, or very flammable. The regulations do not apply to cosmetics and drugs (which are regulated by the Pure Food and Drug Act) and pest control products (regulated by the Pest Control Products Act) (Business Service Center, 2003).

The Canadian Product Safety Program (PSP) assists in the protection of Canadians by researching and assessing the health risks and safety hazards associated with consumer products, cosmetics, and new chemical substances (PSP, 2004). The Consumer Product Safety Bureau (CPS, herein referred to as the Chemistry and Flammability Division) is in charge of this program to promote the safe use of consumer chemical products such as detergents, paints, solvents and glues where the contents may be toxic, corrosive, or flammable. Canada's Cosmetics Program has the mandate to protect the health of Canadians by minimizing the risk associated with the use of cosmetics marketed in Canada. The program defines and communicates requirements for the manufacture, labeling, distribution, and sale of cosmetics. The basis for the regulatory authority for the Cosmetics Program is derived from the Canadian Food and Drugs Act (F&DA) and Cosmetic Regulations (PSP, 2004).

The Environmental Impact Initiative (EII) has been established as a response to growing scientific evidence that substances present in products regulated under the Food and Drugs Act are being found in the environment. EII studies the potential effect of personal care products and pharmaceuticals regulated under the F&DA in the Canadian environment and on human health (EII, 2003). The initiative works collaboratively with the Canadian National Water Research Institute and is involved in developing regulations such as the Canadian Environmental Protection Act (CEPA) and New Substances Notification Regulations (NSNR). Under the directive of EII, Environment Canada has developed the Domestic Substances List (DSL), a comprehensive inventory of known substances in Canadian commerce (past and current). The list currently includes approximately 24,000 substances. The list can be searched online for a specific substance using its CAS Registry Number but gives only information on whether or not the compound is included in the database (Environmental Canada, 2003a). Canada introduced the NSNR in 1994 to ensure that no new substances are introduced on the Canadian market until an assessment has been carried out to determine the risk they pose to the environment and human health. The Domestic Substance List is reviewed by the Existing Substances Branch of Environment Canada to assess environmental impacts of the substances (Environment Canada, 2003b).

The Environmental Impact Initiative's also published a list of product substances and their CAS numbers that were introduced between 1984 and 2001. Production volumes are not available from the online database. When Environment Canada and Health Canada suspect that a Significant New Activity (SNA) in relation to the substance may result in the substance becoming toxic, a notice is issued to ensure that adequate additional information is provided by the notifier or any other proponent who wish to manufacture, import, or use the substance for activities not specified by the notice. To date, the list contains approximately 17 compounds and is accessible online (Environment Canada, 2004).

1.3.1.5 Administration of Household Chemicals in the European Union

The current legislation in the European Union distinguishes between so-called "existing" and "new" chemicals using September 1981 as a cut-off date. New chemicals have to be notified and tested in production volumes as small as 10 kg per year, while there are no such provisions for existing chemicals. The number of new chemicals put on the market since 1981 has reached only around 3,000. The number of "existing" chemicals in 1981 was 100,106. Since 1993, 140 HPV chemicals have been singled out for risk assessment. Only a very limited number has completed the risk assessment process so far.

On 29 October 2003, the EU Commission adopted a proposal for a new EU regulatory framework for chemicals combining all laws regulating chemical-safety issues in a single piece of legislation. The new system is called Registration, Evaluation and Authorisation of Chemicals (REACH). In 2006, REACH became part of a new law for the administration of chemicals (European Commission, 2007). The centerpiece of REACH consists of a registration procedure that is designed to provide systematic information about the properties of relevant substances. The goal of this regulatory framework is to provide information on some 30,000 of the 100,000 existing listed substances, which have an annual production or import volume of over 1 metric ton per manufacturer (approximately 2,200 lb per year). Chemicals that are manufactured or imported in quantities of more than 1 ton per year and per manufacturer/importer will be registered in a central database.

The Joint Research Center (JRC) is currently developing an EU-wide database called EIS-CHEMRISKS to collect and assess information on human exposure to chemicals released from consumer products. This database is not yet available for public use.

HPV chemicals of the European Union are listed in the European Chemical Substances Information System database (ESIS), which currently comprises 2,747 substances with a production or import volume in excess of 1,000 metric ton/yr (European Chemicals Bureau, 2004). The Information System is an extract of data from the IUCLID (International Uniform Chemical Information Data Base) on HPV chemicals reported by European Industry in compliance with the existing European chemicals risk assessment program (European Commission, 1999) (http://ecb.jrc.it). As defined by the European Union, an HPV chemical is a chemical being produced or imported in a quantity of at least 1,000 metric tons (2.2 million lb) per year in the EU by at least one industry. The database lists the chemicals by CAS numbers and provides a list of respective producer and importer companies within the EU. Cross-references are provided to the European Uniform Chemical Information Data Base (EUCLID) database (http://mahbsrv.jrc.it/databases-DBEuclide.html). This database was established to collect, organize, and process data on accidental events involving the unwanted formation of hazardous substances as a consequence of out-of-control conditions in industrial facilities and includes information on toxicity of certain substances. For each substance, the data sheets include summarized data on physicochemical properties, environmental fate and pathways, ecotoxicity and toxicity, and risk assessment.

The National Institute for Public Health and the Environment (RIVM, 1997) attempted to model the environmental fate of HPV chemicals in the environment. Estimates of degradation kinetics for 1,073 compounds in various environmental compartments have been completed. These estimates have been made for all single compounds on the HPV Chemicals (HPVC) list of the European Union (as of June 4th, 1995). Given the large amount of compounds for which no or very few experimental degradation kinetics are available, these estimates can serve as supplementary information for the risk assessment of existing chemicals.

1.3.1.6 Administration of Household Chemicals in Japan

Online information on the administration of household chemicals in Japan is very limited. Japan's law concerning Reporting of Releases of Specific Chemical Substances to the Environment and Promoting Improvements in Their Management requires that businesses handling Class I designated chemicals shall estimate their quantities released to the environment (Ministry of the Environment, 2004).

The Chemical Evaluation and Research Institute (CERI) in Japan maintains a Web-based database on "Biodegradation and bioaccumulation data of existing chemicals." This Website provides experimental data on degradation organized by CAS number for a given chemical (http://qsar.cerij.or.jp/cgi-bin/DEGACC/index.cgi?e).

1.3.1.7 International Initiatives

The Organization of Economic Cooperation and Development (OECD) established the SIDS composed of six hazardous end point tests (i.e., acute toxicity, chronic toxicity, developmental and reproductive toxicity, mutagenicity, ecotoxicity, and environmental fate). These tests do not fully reflect a chemical's toxicity, but they do provide a minimum set of information that can be used to determine the relative hazard of chemicals and to judge if additional testing is necessary (UNEP, 1996).

In 1999, the International Council of Chemical Associations (ICCA) launched the ICCA HPV Chemicals Initiative, a voluntary action by chemical companies, primarily in Europe, Japan, and North America, to deliver globally harmonized data and initial hazard assessments for some 1,300 HPV chemicals. The list of the selected HPV chemicals, including CAS numbers, is available online (ICCA, 2004) (http://www.cefic.org/activities/hse/mgt/hpv/hpvinit.htm). The assessment of the chemicals is sponsored by one of the companies in the listed countries and the report is submitted in form of the SIDS dossier. This list is not a list of all HPV chemicals in the United States. To qualify for the ICCA Working List of HPV substances, chemicals need to be considered HPV or be otherwise of interest in two or more regions (i.e., North America, Europe, or Japan). Chemicals that are HPV in only one of these regions were included if they have been identified as priority chemicals by national and/or regional authorities, if they are chemicals for which assessments in other programs had started but not progressed at early stages, or they fit into a category of chemicals that can be assessed together. Approximately 225 petroleum chemicals were excluded from the ICCA list because they are addressed by other groups (www.cefic.org/activities/hse/mgt/hpv/hpvipe95.xls).

The International Agency for Research on Cancer (IARC), an agency of the World Health Organization (WHO), compiles several databases on carcinogenic risk to humans, epidemiology and cancer control. IARC classifies chemicals and mixtures into four basic groups. Group 1 consists of chemicals that are known to be carcinogenic to humans. Group 2A consists of chemicals that are probably carcinogenic to humans. Group 2B consists of chemicals that are possibly carcinogenic to humans. Group 3 consists of chemicals that are unclassifiable as to carcinogenicity in humans. Group 4 consists of chemicals that are probably not carcinogenic to humans.

CHAPTER 2.0

SELECTION OF HPV TARGET COMPOUNDS

2.1 Selection of HPV Chemicals

2.1.1 Tier Approach

A comprehensive list of household chemicals was initially derived from a database offered by the National Institutes of Health's (NIH) National Library of Medicine (NLM). This was the most comprehensive database identified covering different household commodities. The database is based on material safety data sheets (MSDSs) offered by producers operating within the United States. From this database, approximately 720 compounds were identified that fall within eight commodities: 1) auto products, 2) inside the home, 3) pesticides, 4) home maintenance, 5) personal care/use, 6) pet care, 7) arts and crafts, and 8) landscape/yard. All inorganic and organic compounds were considered from this database even though only organic compounds are targeted in this study. In general, most drugs and food additives were not considered.

The following databases were then used to evaluate production volumes of the 720 identified compounds:

1) International Council of Chemical Associations (ICCA) High Production Volume (HPV) Working List of chemicals which includes chemicals considered HPV by various governments or industries, or otherwise of interest in two or more regions (i.e., North America, Europe, or Japan).

2) U.S. EPA's Toxic Substances Control Act Inventory Update Rule (TSCA IUR) database, which reports on organic chemicals that are manufactured in, or imported into, the United States in amounts equal to or exceeding 10,000 pounds per year. HPV chemicals are defined as those, which are manufactured in or imported into the United States in amounts equal to or greater than one million pounds per year.

3) European Chemical Substances Information System database (ESIS) which reports on chemicals being produced or imported in quantity of at least 1,000 metric tons (equivalent of 2.2 million pounds) per year in the European Union (EU).

The selection of compounds from the comprehensive list of approximately 720 compounds identified for further study and analysis followed a two-tiered ranking approach based on production volumes, environmental relevance, and feasibility for analytical quantification. Originally, the tiered approach was to include three tiers classified as follows:

1) Tier 1: Compounds classified as consumer product chemicals and high release chemicals from the Master Summary Table of the U.S. EPA's HPV Chemical Hazard Data Availability Study.

2) Tier 2: Organic compounds that are either frequently used in households or are present in household commodities.

3) Tier 3: Organic chemicals which are used in households or household commodities, but which are only used in relatively small amounts, such as pharmaceuticals, personal care products, etc.

The three-tiered approach was changed in favor of a two-tiered approach in order to simplify the chemical selection process based primarily on production volumes for those chemicals that are manufactured in or imported into the United States. Additional consideration was given to those chemicals demonstrating positive findings for the six Screening Information Data Set (SIDS) toxicity endpoints (i.e., acute toxicity, chronic toxicity, teratogenicity, mutagenicity, ecotoxicity, and environmental fate) derived from U.S. EPA's Chemical Hazard Data Availability Study.

The selection of Tier 1 chemicals included only those compounds listed in the U.S. EPA's IUR database with yearly production volumes greater than one million pounds per year and which are likely present in domestic wastewater due to their physicochemical properties and reported environmental fate. All Tier 1 chemicals are summarized in the Household Chemical Database in Appendix D (the full database is provided on a CD). The short-list of Tier 1 chemicals targeted in this study for monitoring is listed in Table 2-1 and includes chemicals for which analytical methodologies were established by the research team.

Tier 2 chemicals included compounds with production volumes less than one million pounds per year, or for which no specific production volume was identified, even though these compounds were identified in at least one of the three volume production databases listed above. Tier 2 chemicals are also frequently used in household commodities and are also likely to be present in domestic wastewater due to their physicochemical properties and reported environmental fate. Tier 2 chemicals are compiled in the Household Chemical Database in Appendix D. The short-list of Tier 2 chemicals targeted in this study is listed in Table 2-2 and includes chemicals for which analytical methodologies were established by the research team.

Linear alkylbenzene sulfonate (LAS) and triclocarban were integrated into the target compound list because LAS can function as a model for complex mixtures and triclocarban as a model for emerging contaminants for which only a limited amount of data were available at the time of study initiation (Table 2-3). To study the fate and transport of household chemicals covering a wide range of physicochemical properties during controlled laboratory-scale studies, the team also included several previously studied pharmaceutical residues to the list of compounds of interest. Select physicochemical properties of both the HPV target chemicals and model compounds are summarized in Table 2-4.

Inorganic chemicals identified in the comprehensive database of 720 compounds were not considered in the selection of target compounds for this study. In addition, the small number of compound mixtures identified through one common CAS number (such as certain petroleum products) were also not considered in the selection. Although some organic solvents are listed HPV chemicals, since they have a high tendency to volatize either in the sewer system or during conventional wastewater treatment employing activated sludge systems, these compounds were also not further considered in the selection of target compounds.

2.1.2 Quality Assurance of Databases

It is noteworthy, that the NIH database is established on a *voluntary* disclosure of MSDS information of products by chemical manufacturers. Therefore, not all household products or manufacturers are included in the NIH database. Also, not all product groups are equally well represented in the database; more information is available on U.S. Food and Drug Administration (FDA) regulated products (e.g., cosmetics, shampoos, deodorants, etc.) since more detail is required for labeling (Brad Price, The Procter & Gamble Company, personal communication 2004). Main ingredients of certain product groups (e.g., detergents, cleaners, etc.) are relatively identical among different manufacturers and might be part of the database even if certain manufacturers are not participating.

Evaluating how comprehensively the NIH database covers household chemicals and which chemical volumes derive specifically from household chemicals (and not from other industrial sources) is not an easy task. Several promising approaches (market research [i.e., MindBranch or the Chemical Economics Handbook] and estimating shelf space in stores, etc.) were not feasible considering time and financial resources available through this study. Instead, we attempted to compare the NIH database to other sources accessible online which list common household chemicals. Findings of this comparison are reported for certain product lines below.

- Huish Detergents, Inc. is a fast growing maker of store brand detergents. Nevertheless, the company's products are not included in the NIH database. The researchers compared their database with the compounds listed on all MSDS sheets provided on the company's webpage (http://www.huish.com/msd_sheets.html). The MSDS sheets only reveal hazardous ingredients. Of all organic compounds listed (17 total for all products), 10 were included in our database. Of the seven compounds not included, none were HPV chemicals according to ICCA or IUCLID. Only one compound (i.e. pine oil, a solvent) was a HPV according to U.S. EPA's Inventory Update Rule (IUR) listing.
- Redgum Soaps (http://www.redgumsoaps.com.au/toxic_chemicals.().html) provides a list of the more common ingredients used in skincare products. The list includes 42 organic compounds out of which 39 were included in the NIH database.
- U.S. EPA maintains a website entitled "Hazardous Products in the Home" (http://www.epa.gov/grtlakes/seahome/housewaste/house/mainmenu.htm) that features particularly hazardous chemicals for direct exposure that can occur in different parts of the house (e.g., bath, kitchen, and garage). Of the 46 organic compounds listed, 35 were included in the NIH database. Of the 11 compounds not listed, five were organic solvents. Overall the NIH database covered 87% of all HPV and non-HPV chemicals in the U.S. EPA listing.
- A similar website is provided by Meyer Groundwater Services (http://meyergs.com/default.pk?tsearch=Household+hazards&search_button.x=0&search_button.y=0), offering information on hazardous chemicals in household products. The majority of the 50 organic compounds listed overlaps with the compounds on U.S. EPA's webpage. A total of five compounds (10%) were not listed in the NIH database.

Although limited, these comparisons indicate the NIH list does include the most common organic household products in the U.S. that are HPV and known hazards (approximately 90%).

The database certainly does not include all ingredients in household products since there are probably thousands of them. In Europe efforts are being made to account for all the ingredients, for example, European regulations for cosmetic products require a long list of ingredients in individual product lines. Since 1998, ingredients must be approved and included in the so-called INCI list (International Nomenclature of Cosmetic Ingredients) before use. The list alone comprises more than 400 mostly organic compounds or compound groups, including a large group of pharmaceuticals.

An additional quality assurance measure consists of comparing the list of selected target compounds with chromatograms derived from gas chromatography with mass spectrometry (GC/MS) for field samples collected by members of the research team from wastewater dominated streams. Previously undetected compounds were identified using the National Institute of Standards and Technology (NIST) mass spectral library. Structural confirmation is achieved in this approach using authenticated standards, when available.

Table 2-1. Selected Tier 1 Household Chemicals.

Compound	Chemical Abstracts Service #	Hazard Data Value[1]	Synonyms	Applications	Analytical Method
2,6-Di-t-butyl-p-cresol (BHT)	000128-37-0	6	2,6-Di-t-butyl-p-cresol; BHT; 2,6-Di-tert-butyl-p-cresol; Butylated hydroxytoluene; Butylhydroxytoluene; Di-t-butylcresol; Phenol, 2,6-bis(1,1-dimethylethyl)-4-methyl-; p-Cresol, 2,6-di-tert-butyl-	skin care, hobby supplies	GC/MS-MS
Dibutyl phthalate	000084-74-2	6	1,2-Benzenedicarboxylic acid, dibutyl ester; Butyl phthalate; DBP; Di-n-butyl phthalate; Dibutyl 1,2-benzenedicarboxylate; Dibutyl phthalate; Phthalic acid, dibutyl ester	plasticizer, finish, nail care	GC/MS-MS
Atrazine	001912-24-9	6	2-Chloro-4-(2-propylamino)-6-ethylamino-s-triazine; 2-Chloro-4-ethylamino-6-isopropylamino-s-triazine; 2-Chloro-4-ethylamino-6-isopropylamino-1,3,5-triazine; 2-Chloro-4-ethylamineisopropylamine-s-triazine; 2-Ethylamino-4-isopropylamino-6-chloro-s-triazine	herbicide	LC/MS-MS
4,4-Isopropylidenediphenol (Bisphenol A)	000080-05-7	5	4,4'-(1-Methylethylidene)bisphenol; 4,4'-Bisphenol A; 4,4'-Isopropylidenediphenol; Bisphenol A; Diphenylolpropane; Phenol, 4,4'-(1-methylethylidene)bis-; Phenol, 4,4'-isopropylidenedi-; 4,4' Isopropylidinediphenol	plasticizer, epoxy, glue	LC/MS-MS
Benzophenone	000119-61-9	4	Benzophenone; Benzoylbenzene; Diphenyl ketone; Diphenylmethanone; Methanone, diphenyl-	car wash, facial products	GC/MS-MS
Benzophenone-3 (Oxybenzone)	000131-57-7	3	(2-Hydroxy-4-methoxyphenyl)phenylmethanone; 2-Hydroxy-4-methoxybenzophenone; Benzophenone-3; Methanone, (2-hydroxy-4-methoxyphenyl)phenyl-; Oxybenzone; Benzophenone, 2-hydroxy-4-methoxy-	skin care, insect repellant	LC/MS-MS
Triclosan	003380-34-5	n.i.	Triclosan; Phenol, 5-chloro-2-(2,4-dichlorophenoxy)-; Ether, 2'-hydroxy-2,4,4'-trichlorodiphenyl; 5-Chloro-2-(2,4-dichlorophenoxy)phenol; 2,4,4'-Trichloro-2'-hydroxy diphenyl ether	antibacterial, soap, lotion	LC/MS-MS
Vanillin	000121-33-5	6	p-Hydroxy-m-methoxybenzaldehyde; 4-Hydroxy-m-anisaldehyde; 4-Hydroxy-3-methoxybenzaldehyde; Benzaldehyde, 4-hydroxy-3-methoxy-; Vanillin [USAN]	fragrance, cosmetics, various	GC/MS-MS
o-Phenylphenol	000090-43-7	6	(1,1'-Biphenyl)-2-ol; 1-Hydroxy-2-phenylbenzene; 2-Phenyl phenol; Biphenyl-2-ol; 2-Phenylphenol; Biphenyl, 2-hydroxy-; o-Phenylphenol	germicide, fungicide	GC/MS-MS LC/MS-MS
2-Phenoxyethanol	000122-99-6	5	Phenoxyethanol; 2-Phenoxyethanol; Ethanol, 2-phenoxy-; Ethylene glycol monophenyl ether; Phenoxytol	preservatives, cosmetics, various	GC/MS-MS
Hexabromocyclododecane	003194-55-6	4	1,2,5,6,9,10-Hexabromocyclododecane; Hexabromocyclododecane; Cyclododecane, 1,2,5,6,9,10-hexabromo-	flame retardant	GC/MS-MS

Notes:

1 – Derived from EPA's Chemical Hazard Data Availability Study. Value reflects testing conducted relevant to the six Screening Information Data Sets (SIDS) toxicity endpoints (acute toxicity, chronic toxicity, teratogenicity, mutagenicity, ecotoxicity, and environmental fate).

n.i. – Chemical not found within SIDS database and/or U.S. EPA's TSCA IUR database.

Table 2-2. Selected Tier 2 Household Chemicals.

Compound	Chemical Abstracts Service #	Hazard Data Value[1]	Synonyms	Applications	Analytical Method
Simazine	000122-34-9	6	1,3,5-Triazine-2,4-diamine, 6-chloro-N,N'-diethyl-; s-Triazine, 2-chloro-4,6-bis(ethylamino)-; Simazine	herbicide, anti-algae products	LC/MS-MS
N,N-Diethyl-meta-toluamide (DEET)	000134-62-3	4	N,N-Diethyl-meta-toluamide; DEET; Benzamide, N,N-diethyl-3-methyl-; N,N-Diethyl-m-toluamide; m-Toluamide, N,N-diethyl-	insect repellant, industrial	LC/MS-MS GC/MS-MS
Hydrocortisone	000050-23-7	n.i.	11-beta-Hydrocortisone; 11beta,17,21-Trihydroxypregn-4-ene-3,20-dione; 17-Hydroxycorticosterone; 17alpha-Hydroxycorticosterone; 4-Pregnene-11beta,17alpha,21-triol-3,20-dione; Cortisol; Hydrocortisone; Hydroxycortisone	anti-itch cream	LC/MS-MS
Butylated hydroxyanisole (BHA)	025013-16-5	n.i.	BHA; tert-Butyl-4-methoxyphenol; 3-tert-Butyl-4-hydroxyanisole; Anisole, butylated ydroxyl-; Phenol, (1,1-dimethylethyl)-4-methoxy-	anti-oxidant, various	GC/MS-MS LC/MS-MS
3-Indolebutyric acid	000133-32-4	n.i.	4-Indol-3-ylbutyric acid [BSI:ISO]; Indole-3-butyric acid; Butyric acid, 4-(indolyl)-; Indole butyric acid; Indole-3-butanoic acid	fertilizer, miracle gro™	LC/MS-MS
Camphor	76-22-2	n.i.	2-Bornanone; 1,7,7-Trimethylbicyclo(2.2.1)heptan-2-one; Bicyclo(2.2.1)heptan-2-one, 1,7,7-trimethyl-; Camphor; Spirit of camphor; Gum camphor; (+-)-Camphor; DL-Bornan-2-one	fragrance, various	GC/MS-MS
Menthol	001490-04-6	n.i.	Menthol; 5-Methyl-2-(1-methylethyl)cyclohexanol; 2-Isopropyl-5-methylcyclohexanol; Menthyl alcohol; Cyclohexanol, 5-methyl-2-(1-methylethyl)-; DL-Menthol	fragrance, various	GC/MS-MS
2-Methylresorcinol	000608-25-3	n.i.	1,3-Benzenediol, 2-methyl-; Resorcinol, 2-methyl-	hair color, cosmetics	GC/MS-MS
Isobutylparaben	004247-02-3	n.i.	Benzoic acid, p-hydroxy-, isobutyl ester; Benzoic acid, 4-hydroxy-, 2-methylpropyl ester (9CI); Isobutyl 4-hydroxybenzoate; 2-Methylpropyl 4-hydroxybenzoate; Isobutylparaben; Isobutyl parahydroxybenzoate; 4-Hydroxybenzoic acid, 2-methylpropyl ester; Isobu	preservative, cosmetics, various	LC/MS-MS
Acriflavine	008048-52-0	n.i.	Xanthacridinum; Acridinium, 3,6-diamino-10-methyl-, chloride, mixt. With 3,6-acridinediamine	pesticide	LC/MS-MS
Trifluralin	001582-09-8	n.i.	2,6-Dinitro-N,N-di-n-propyl-alpha,alpha,alpha-trifluoro-p-toluidine; Trifluralin; Benzenamine, 2,6-dinitro-N,N-dipropyl-4-(trifluoromethyl)-; N,N-Dipropyl-2,6-dinitro-4-trifluoromethylaniline; 4-(Trifluoromethyl)-2,6-dinitro-N,N-dipropylaniline; 2,6-Dinit	herbicide	GC/MS-MS
2,3,4,5 Bis(2-butylene) tetrahydro 2-furaldehyde (MGK-11)	000126-15-8	n.i.	Dibutylene tetrafurfural; 2,3:4,5-Di(2-butenyl)tetrahydrofurfural; 1,4,4°,5°,6,9,9°,9b-Octahydrodibenzofuran-4°-carbaldehyde	insect repellant	GC/MS-MS
Propylparaben	000094-13-3	n.i.	4-Hydroxybenzoic acid, propyl ester; Benzoic acid, 4-hydroxy-, propyl ester; Benzoic acid, p-hydroxy-, propyl ester; Propyl 4-hydroxybenzoate; Propyl parahydroxybenzoate; Propyl p-hydroxybenzoate; Propyl paraben	preservative, various	LC/MS-MS

Notes:
1 - Derived from EPA's Chemical Hazard Data Availability Study. Value reflects testing conducted relevant to the six Screening Information Data Sets (SIDS) toxicity endpoints (acute toxicity, chronic toxicity, teratogenicity, mutagenicity, ecotoxicity, and environmental fate).
n.i. - Chemical not found within SIDS database and/or U.S. EPA's TSCA IUR database.

Table 2-3. Selected Model Compounds.

Compound	CAS #	Hazard Data Value[1]	Synonyms	Applications	Analytical Method
Triclocarban	000101-20-2	6	1-(3',4'-dichlorophenyl)-3-(4'-chlorophenyl)urea; 3,4,4'-trichlorodiphenylurea; 3,4,4'-Trichlorocarbanilide; cusiter; cutisan; genoface; N-(3,4-dichlorophenyl)-N'-(4-chlorophenyl)urea; N-(4-chlorophenyl)-N'-(3,4-dichlorophenyl)urea; procutene; Solubacter; TCC; Trichlorocarbanilide; triclocarban	antibacterial, soap, deodorant	LC/MS-MS
Linear alkylbenzene sulfonate (LAS)	042615-29-2	n.i.	n.a.	surfactant, various	LC/MS

Notes:

1 - Derived from U.S. EPA's Chemical Hazard Data Availability Study. Value reflects testing conducted relevant to the six Screening Information Data Sets (SIDS) toxicity endpoints (acute toxicity, chronic toxicity, teratogenicity, mutagenicity, ecotoxicity, and environmental fate).

n.i. - Chemical not found within SIDS database and/or U.S. EPA's TSCA IUR database.

n.a. - information not available.

Table 2-4. Physicochemical Properties of Target Compounds.

Compound	Mol. Wt.[1, 2] (g/mol)	Henry's Constant[1, 2] (atm-m^3/mol)	Water Solubility [1, 2] (mg/L)	pKa[1, 2]	log K_{ow} [1, 2] (octanol-water)
Indicator Compounds					
Primidone	218.25	1.94E-10	500	11.3/12.25	0.91
Phenacetine	179.22	2.13E.10	766	14.13	1.58
Carbamazepine	236.28	1.08E-10	17.66	0.37 (base)	2.45
2-Naphthol	144.17	2.74E-08	755	9.51	2.7
Fenofibrate	360.8	4.46E-09	0.418	n.a.	5.19
Gemfibrozil	250.3	1.19E-08	10.9	4.75	0.79
Propyphenazone	230.3	1.84E-09	3.0E+06	0.11/-4.92/6.16	1..94
Sulfamethoxazole	253.4	6.93E-08	610	1.97/0.24 (base)	-0.07
Ibuprofen	206.28	1.50E-07	21	4.91	-0.4
Naproxen	230.26	3.39E-10	159	4.15	3.18
Diclofenac	296.16	4.73E-12	2.37	4.15	4.51
Range	**128.2 - 296.16**	**4.73E-12 - 4.40E-4**	**2.37 - 766**	**4.15 - 9.51**	**-0.4 – 5.19**
Average	**204.85**	**6.29E-05**	**281.50**	**5.68**	**2.28**
Tier 1 Compounds					
2,6-Di-t-butyl-p-cresol (BHT)	220.36	4.12E-6	0.6	12.2	5.10
Dibutyl phthalate	278.35	1.81E-06	11.2	n.a.	4.5
Atrazine	215.69	2.36E-09	34.7	1.7	2.61
4,4-Isopropylidenediphenol (Bisphenol A)	228.29	1.00E-11	120	n.a.	3.32
Benzophenone	182.22	1.94E-6	137	n.a.	3.18
Benzophenone-3 (Oxybenzone)	228.25	1.50E-08	68.6	n.a.	3.79
Triclosan	289.55	4.99E-09	10	n.a.	4.76
Vanillin	152.15	2.15E-09	11,000	7.4	1.21
o-Phenylphenol	170.21	1.05E-06	700	9.97	3.09
2-Phenoxyethanol	138.17	4.72E-08	26,700	15.1	1.16
Hexabromocyclododecane	641.7	4.60E-05	0.0086	n.a.	7.74
Range	**138.17 – 641.7**	**1.00E-11 – 4.60E-05**	**0.0086 – 26,700**	**1.7 – 15.1**	**1.16 – 7.74**
Average	**249.54**	**5.00E-06**	**3,530**	**9.27**	**3.68**

Tier 2 Compounds					
Simazine	201.66	9.42E-10	6.2	1.62	2.18
N,N-Diethyl-meta toluamide (DEET)	191.28	2.08E-08	912	n.a.	2.18
Hydrocortisone	362.47	5.77E-08	320	n.a.	1.61
Butylated hydroxyanisole (BHA)	180.3	1.17E-06	213	n.a.	3.5
3-Indolebutyric acid	203.24	1.28E-11	250	4.7	2.3
Camphor	152.24	8.10E-05	1,600	n.a.	2.38
Menthol	156.27	1.52E-05	420	n.a.	3.38
2-Methylresorcinol	124.14	6.44E-11	263,000	n.a.	1.58
Isobutylparaben	194.23	8.45E-09	224	n.a.	3.4
Acriflavine	259.74	1.79E-19	250,000	n.a.	-2.64
Trifluralin	335.29	1.03E-04	0.184	n.a.	5.34
2,3,4,5 Bis(2-butylene) tetrahydro 2-furaldehyde (MGK-11)	204.27	7.98E-07	8,130	n.a.	1.51
Propylparaben	180.21	6.37E-09	500	n.a.	3.04
Range	**124.15 – 362.47**	**1.79E-19 – 1.03E-04**	**0.184 – 263,000**	**1.62 – 4.7**	**-2.64 – 5.34**
Average	**225.04**	**1.55E-05**	**40,429**	**3.16**	**2.29**

Compound	**Mol. Wt.**[1, 2] (g/mol)	**Henry's Constant**[1, 2] (atm-m^3/mol)	**Water Solubility**[1, 2] (mg/L)	**pKa**[1, 2]	**log K_{ow}**[1, 2] (octanol-water)
		Model Compounds			
Triclocarban	315.59	4.5E-11	0.0237	n.a.	4.9
Linear alkylbenzene sulfonate (LAS)	n.a. - various	n.a. - various	n.a. - various	n.a. - various	n.a. – various

Notes: 1 – Values obtained from Syracuse Research Corporation at http://www.syrres.com/esc/physprop.htm.

2 – Values obtained from http://www.chemfinder.com.

n.a. – Value not available.

CHAPTER 3.0

ANALYTICAL METHODS AND DESIGN OF FIELD-SCALE MONITORING PROGRAM

3.1 Chemicals and Materials

All chemicals used were of reagent grade or higher, purchased from J.T. Baker, Inc. (Philippsburg, NJ), CDN Isotopes (Quebec, Canada) Sigma-Aldrich (St. Louis, MO), Eastman Organic Chemicals (Rochester, NY), and Fisher Scientific (Fairlawn, NJ).

3.2 Bulk Measurements

3.2.1 General Parameter

3.2.1.1 Conductivity

Conductivity is an estimate of the total dissolved solids content of water, and is measured in microSiemens per centimeter (μS/cm). Conductivity measurements were taken with a handheld Cole Parmer EC Meter (model 1481-61) (Standard Method 2510). Prior to use, the conductivity meter was calibrated with a 1,000 micromho per centimeter conductivity/TDS calibration solution obtained from RICCA Chemical Company.

3.2.1.2 pH

The pH was determined using a Beckman 260 portable pH meter with combination of a gel-filled electrode (Beckman, Fullerton, CA) (Standard Method 4500-H^+). Prior to use, the pH meter was calibrated with calibration solutions of pH 4, 7, 10 obtained from RICCA Chemical Company.

3.2.1.3 Total Organic Carbon

TOC is a measure of the concentration of total organic carbon in a water sample. For dissolved organic carbon (DOC), the sample was first filtered using a 0.45 μm filter to remove particulate organic carbon. Organic carbon concentrations were determined at CSM using a Sievers 800 Total Organic Carbon Analyzer (Ionics, Boulder, CO) according to Standard Method 5310 B. The Sievers 800 measurements are based on the oxidation of organic carbon to form carbon dioxide using ammonium persulfate and UV radiation. Carbon dioxide is measured using a selective membrane-based conductometric detection technique. The detection limit for TOC using the Sievers 800 was 0.06 mg/L. Fresh standard solutions were run with samples for each measurement.

3.2.1.4 UV Absorbance

UVA is a measure of the quantity of compounds present in a sample that absorb ultra-violet (UV) light at a wavelength of 254 nm. Compounds that absorb UV light at 254 nm are usually aromatic in structure. UVA measurements at CSM were conducted with a Beckman DU800 UV/Vis Scanning Spectrophotometer (Fullerton, CA) according to Standard Method

5910 B. The absorbance was first zeroed using ultra-pure water created by processing in house deionized water through a Millipore Milli-Q™ water system. Cuvettes used are 1 cm and 4 cm, rectangular, quartz glass cuvettes. Once the absorbance was zeroed, the absorbance of each sample is read twice, and the average of the two readings is recorded. The specific UV absorbance (SUVA) was calculated as the ratio between UV absorbance and DOC.

3.2.1.5 Ammonia, Nitrite, Nitrate and Phosphorus

Ammonia was measured using a Hach DR/2500 spectrophotometer at CSM (Hach method 8038). Nitrite, nitrate and phosphorus (as o-phosphate) were measured with a Dionex DS600 Ion Chromatograph (IC) (Dionex, Sunnyvale, CA) using an AS14 column and a sodium hydroxide eluent according to Standard Method 4110C. Samples were 0.45 µm-filtered prior to analysis.

3.2.1.6 Other Inorganic Ions

With the exception of ammonium, inorganic cations were measured at CSM with an Optima 3000 Inductive Coupled Plasma (ICP) Spectrometer (Perkin Elmer, Norwalk, CT) according to Standard Method 3120B. Prior to analysis, samples were pretreated by 0.45 µm membrane filtration and acidified to pH 2.

3.3 Extraction Methods for Target Compounds

Based on the results of method development efforts for the original 37 proposed target compounds, the researchers were able to develop LC/MS-MS and GC/MS-MS methods for 26 of the compounds, including triclocarban. LAS compounds were analyzed in split samples by Dr. Brad Price, The Procter & Gamble Company. The target compounds are summarized in Tables 2-1, 2-2, and 2-3.

3.3.1 Extraction for Liquid Chromatography with Tandem Mass Spectroscopy

Analytes were extracted in batches of six samples using 200-mg hydrophilic-lipophilic balance (HLB) glass cartridges from Waters Corp. All extractions were performed on an AutoTrace automated SPE system. The SPE cartridges were sequentially preconditioned with 5 mL MTBE, 5 mL methanol, and 5 mL reagent water. Depending on the wastewater content of the sample, between 100 mL and 1000 mL of sample were used. After spiking with surrogates, a sample was loaded onto the cartridges at 15 mL/min after which the cartridges were rinsed with 5 mL reagent water and then dried with a stream of nitrogen for 60 minutes. Next, the cartridges were eluted with 5 mL of 10%/90% (v/v) methanol/MTBE followed by 5 mL of methanol into 15-mL calibrated centrifuge tubes. The resulting extract was concentrated with a gentle stream of nitrogen to a volume of ~400 µl. The extract was then brought to a final volume of 500 µl with methanol.

3.3.2 Extraction for Gas Chromatography with Mass Spectroscopy

Samples were extracted in batches of six using 200-mg hydrophilic-lipophilic balance (HLB) glass cartridges from Waters Corp. All extractions were performed using an Autotrace automated SPE system. The HLB cartridges were sequentially preconditioned with 5 mL dichloromethane, 5 mL MTBE, 5 mL methanol, and 5 mL reagent water. Five hundred-mL (500-mL) samples were spiked with 250 µl of a surrogate standard solution containing 1 mg/L of [$^{13}C_6$]-o-phenylphenol, BHT-d_{24}, dibutyl phthalate-d_4 and 4 mg/L of [$^{13}C_6$]-vanillin. For influent wastewater, only 100 mL of sample was used for extraction, which was diluted into DI water for a final volume of 500 mL. The samples were then loaded onto SPE cartridges at 15 mL/min. After sample loading, the SPE cartridges were rinsed with 5 mL reagent water and dried with a stream of nitrogen for 30 min. The SPE cartridges were eluted with 5 mL of 10%/90% (v/v) methanol/MTBE followed by 5 mL of methylene chloride. The eluted extract was collected in a 15-mL calibrated glass vial and concentrated with a gentle stream of nitrogen to 5 mL. The extract was dried over sodium sulfate and transferred to another Zymark vial. The remaining sodium sulfate was rinsed twice with 1 mL of methylene chloride, which was decanted off and combined with the sample extract. The sample was concentrated to 2.0 mL with a gentle stream of nitrogen. At this point, 1 mL of iso-octane was added and the extract was further concentrated to just below 0.5 mL, at which time 25 μl of a 5 mg/L solution of internal standards ($[^{13}C_6]$-α-BHC, $[^{13}C_{12}]$-methoxychlor, DDD-d_8) was added and brought to a final volume of 0.5 mL.

3.4 Chemical Analyses

3.4.1 Liquid Chromatography-Tandem Mass Spectrometry

Mass spectrometry was performed using an API 4000 triple quadrupole mass spectrometer (Applied Biosystems, Foster City, CA). Due to differing physical properties of the target study compounds, two LC methods were developed. One of the methods used a formic acid eluent and relies on electrospray ionization (ESI) in the positive mode. The acidic eluent served to promote proton adducts during ionization for compounds that were amenable to this process such as bases. The other method used an ammonium acetate eluent and ionization occurs in ESI negative mode. This eluent allowed the formation of negative ions and worked well for compounds such as acids.

3.4.1.1 ESI Positive Method

This method used a Phenomenex Synergi Max RP reverse phase column (4.6 mm x 250 mm) and a dual eluent system composed of 0.1% formic acid in water (A) and 100% methanol (B). The flow rate was 700 μl/min and the gradient was as follows: 5% B held for 0.5 minutes, increased linearly to 100% B by 8.5 minutes, and held at 100% B for 6.5 minutes. After each run, the column was equilibrated for 9.0 minutes with 5% B. This resulted in a total analysis time of 24 minutes. A volume of 35 μl of sample was injected into a 10 μl sample loop. All ESI positive compound-dependent parameters can be found in Appendix B, Table B-2.

3.4.1.2 ESI Negative Method

This method used a Phenomenex Synergi Max RP reverse phase column (4.6 mm x 250 mm) and a dual eluent system composed of 100-mM ammonium acetate in water (A) and 100% methanol (B). The flow rate was 700 μL/min and the gradient was as follows: 5% B held for 0.5 minutes, increased linearly to 100% B by 8.5 minutes, and held at 100% B for 8.5 minutes. After each run, the column was equilibrated for 9.0 minutes with 5% B. This resulted in a total analysis time of 26 minutes. A volume of 35 μl of sample was injected into a 10-μl sample loop. All ESI negative compound-dependant parameters are summarized in Appendix A, Table A-3. Source-dependent parameters for both methods can be found in Appendix A, Table A-4.

3.4.1.3 Recoveries and Calibration

Method reporting limits and average compound recovery data are provided in Appendix B, Table B-5. Calibration curves contained at least five points and the lowest calibration point was the same or lower than the method reporting limit. Fits were either linear or quadratic and correlation coefficients were at least 0.995 or higher. After the instrument calibration, up to six samples were analyzed followed by a calibration check. Depending on the nature of the samples, fewer samples may be run between calibration checks. High concentrations of wastewater have been seen to significantly degrade instrument performance, necessitating more frequent instrument cleaning and calibration.

Surrogate standards were added to all samples to verify the quality of the individual sample extraction. With each batch of sample extractions, an extracted blank and an extracted fortified blank were run to assess the quality of the sample extractions and detect any extraction problems.

3.4.1.4 LC/MS-MS Compound Selection Summary

A list of 40 compounds was selected for initial testing using LC/MS-MS. All compounds were infused into the mass spectrometer to determine whether ionization was possible and to what degree. A summary of the results of this process and reasons for the exclusion of compounds are reported in Appendix B, Table B-1. After infusion, 29 compounds were found to be acceptable. Next, ionization source-dependent parameters were investigated using a mixture of all 29 compounds and values were set that resulted in the highest sensitivity and selectivity for most of the compounds. At this time, LC methods for both ionization modes were developed and initial chromatography parameters were selected. As a result of this testing, 17 compounds were found to remain acceptable. Solid phase extractions (SPE) were then tested and this resulted in the loss of 3 more compounds, yielding the final total of 14 compounds. After these compounds were finalized, representative isotopically labeled surrogate standards were obtained and used to monitor extractions and for quality control purposes. These were carbon-13 labeled simazine, atrazine, bisphenol A, o-phenylphenol, and triclosan.

3.4.2 Gas Chromatography-Mass Spectrometry

A Varian CP-3800 Gas Chromatograph with a CP-8400 autosampler was used for all analyses. The injector (Varian 1177) was operated in splitless mode with a Siltek deactivated glass liner with glass frit. Analytes were separated on a 30 m x 0.25 mm ID x 0.25 µm DB5-MS column (J & W, Agilent) using a 1.0-mL/min helium flow with an initial pressure pulse of 45 psi for 0.85 min. The temperature program employed was as follows: 90°C, hold for 2.0 min; 90-280°C at 7°C/min, hold for 2.0 min; 280°to 315°C @ 50°C/min, hold for 5.16 min. An injection volume of 2 µl was used for all analyses.

3.4.2.1 Mass Spectrometry

For analyses, mass spectrometry was performed using a Varian 2200 ion trap mass spectrometer. All analyses were performed using multiple reaction monitoring (MRM) in positive electron impact mode. For each compound, two mass transitions were used. The first transition is the most abundant and used for quantification of the analyte. The second was used for confirmation, to further validate that the identification was correct. Mass transitions are listed in Appendix C, Table C-1. Vanillin and benzophenone did not have a second mass transition available, as the mass transition chosen for quantization was the only one possible at these ppb levels.

3.4.2.2 Recoveries and Calibration

Expected method reporting limits (MRL) and average percent recoveries are listed in Appendix C, Table C-2. Recoveries were tested in DI, raw intake, and Las Vegas Wash (Wash) water matrices. Reported recoveries were an adjusted average, in which the environmental levels found in the unspiked samples were subtracted out. These blank values are listed in Appendix C, Table C-3 for unspiked DI, raw, and wash samples. Reporting limits can change depending on levels of compounds that appear in blanks. As many of these are common in everyday products, many are being detected in extracted and instrument blanks. For example, dibutyl phthalate is present in GC injector septa. This contributed to instrument blank contamination, although the levels were much lower than what has been seen in the environment. Another example is benzophenone, which is used in bath and skin care products. Contamination is most likely from the sample collection and extraction process from the people working with the samples. The method reporting limits of 25, 50, and 100 ng/L were set at the lowest calibration point, chosen for a signal to noise ratio greater than 10 for all compounds. Analyte quantification was performed using internal standard calibration with linear or quadratic regression, depending upon the analyte response. Correlation coefficients must be consistently 0.990 or greater. During the course of sample analysis, a quality control check was done every six samples to ensure that the performance of the instrument had not altered. If the quality control check did not agree with the original standard within a +/-20% window, the samples were stopped and rerun.

Recoveries in spiked wash samples appeared to have higher recoveries than the DI and raw water spiked samples. This is mostly due to the fact that the target compounds are present in the unspiked samples but are below the reporting limit and, therefore, no recovery adjustment is done. Another reason is that it is assumed when the average recovery is adjusted for detected blank levels, that the spiked samples have the same amount as the unspiked samples. Because of the wide and varied blank contamination of dibutyl phthalate, the recoveries of this compound will be more sporadic resulting is larger relative standard deviations (RSD) as compared to the other compounds.

During method development a second elution was tested using 10 mL of methylene chloride to determine the best eluent for sample extractions. This was based on SPE tests that were done prior to this project testing a wide variety of compounds. These previous tests were done using multiple variations of eluent with a 10%/90% methanol/MTBE resulting in the highest recoveries. Therefore this elution was tested with the target compounds chosen for this project. Also tested for the chosen compound list were an elution with only methylene chloride and an elution with methanol, MTBE, and methylene chloride which involved a micro liquid/liquid extraction. These results are compared in Appendix C, Table C-4. With the exception of vanillin, all compound recoveries were improved and m-resorcinol was recovered from zero recovery with the other SPE methods.

3.5 Liquid Chromatography-Mass Spectrometry for Linear Alkylbenzene Sulfonate (LAS)

Whole wastewater and effluent samples were preserved in the field with 1% (v/v) 37% formalin and shipped to the Procter & Gamble laboratory for analysis. Aliquots of raw wastewater (10–20 mL) were diluted tenfold prior to concentration/analysis to minimize matrix effects. Samples of treated wastewater effluents (200–600 mL) were analyzed directly.

Samples (1:10 raw wastewater or 100% treated effluent) were spiked with approximately one microgram of d3-c12 LAS internal standard, concentrated using a C2 solid phase extraction cartridge (Varian BondElut C2 1gm/6cc), eluted with methanol, evaporated to dryness under nitrogen and reconstituted to 1mL with 1:1 methanol:water. Instrumental analysis uses a high-performance liquid chromatography with negative ion electrospray mass spectrometry detection (Waters Alliance/ZMD single quadropole MS) and internal standard quantitation.

3.6 Liquid Chromatography with UV Detection for Target Compounds Used in Laboratory-scale Studies

Select target (indicator) compounds used in laboratory-scale spiking experiments (Table 2.4) were quantified in the ppb-range using a Hewlett Packard 1100 Series High Performance Liquid Chromatograph with UV detection (HPLC-UV) (Agilent, Palo Alto, CA) with a Supelco C-18 (Sigma-Aldrich, St. Louis MO) reversed phase column. The samples were injected directly without any additional preparation step. A summary of the operational conditions of the HPLC method is presented in Table 3-1.

Table 3-1. Operational Conditions of HPLC-UV Method for Select Target Compounds.

Solvent A	25 mMol KH_2PO_4 DI water adjusted to pH = 2.5 with HPLC grade o-phosphoric Acid
Solvent B	HPLC grade acetonitrile with the same concentration of o-phosphoric acid as Solvent A
Gradient	t = 0 min 20% t = 14 min 80% t = 17 min 30% Post-run time = 4 minutes
UV wavelength program	t = 0 min 205 bw = 10, reference = 330 bw = 60 t = 8.5 min 230 bw = 10, reference = 330 bw = 60 t = 9.5 min 205 bw=10, reference = 330 bw=60
Injection volume	100 µL
Flow rate	2.0 mL/min

The mobile phase consisted of acetonitrile and ultra-pure water, both adjusted to pH 2.5. In reversed phase HPLC, compounds are retarded by hydrophobic interactions to the solid phase. For better resolution, a gradient program was executed with a linearly increasing percentage of acetonitrile in the mobile phase resulting in decreasing polarity. To retard hydrophilic compounds, especially primidone, a low concentration of acetonitrile was required. A high percentage of acetonitrile is necessary to remove hydrophobic compounds from the solid phase. For detection of the compounds an ultraviolet diode array detector (UV-DAD) was used. The detector works by measuring a certain wavelength compared to a reference wavelength. A wavelength program was used to ensure best detection of all compounds. The reference wavelength and bandwidth were selected so that the absorbance due to the compounds was negligible. This was experimentally found to be 330 nm with a bandwidth of 60 nm. Standards were measured in the range of 10 to 500 µg/L. The limit of quantification (LOQ) for all compounds was 10 µg/L. For ibuprofen the LOQ was 20 µg/L.

3.7 Field Sites Selected for Monitoring

Field investigations for this study were carried out at six participating facilities located in three different states within the western continental United States. This approach allowed an investigation of both regional variations of effluent quality and the removal effectiveness of individual processes comprising an overall process train. During the initial phase of the study, the research team conducted a survey at all participating utilities to determine unit processes and operations employed, process parameters, water qualities and service area characteristics for each individual wastewater treatment plant participating in this study. The survey represented the basis for selecting field sites, unit operations, and sampling logistics for full-scale sampling. Table 3-2 summarizes the treatment specifics and sampling locations of each plant selected for this study. The facility summary presented in Appendix A describes the unit processes and operations employed for each of the facilities surveyed, as well as service area characteristics, capacity, influent and effluent water qualities, sludge handling and treatment, and type of discharge. The participating utilities assisted with sampling logistics and provided operational information and data for the time of sampling. A summary of key operational characteristics of the selected full-scale installations is presented in Table 3-3.

3.8 Sample Collection, Preparation, and Storage

Each facility was instructed to follow sampling guidelines provided by the research team. Efforts were made to reduce, if not eliminate, the use of plastics during the sampling process because they can either cause adsorptive losses of target compounds or leach compounds targeted in the study into the sample.

3.8.1 Sampling Equipment

The bottles used for sampling were pre-cleaned 1-L amber glass bottles (Eagle-Picher, Phoenix, AZ) provided by the Southern Nevada Water Authority (SNWA) to each of the facilities. Additional sampling equipment was rinsed with soap and water, hydrochloric or nitric acid, methanol, and finally rinsed with ultra-pure water.

3.8.2 Sample Collection

During field-scale occurrence monitoring samples were collected over a 4-hour or 24-hour period. Prior to sampling, the proper operation of the plant was confirmed by operational parameters provided by the facility in the survey. Care was also taken not to sample the facilities within 48 hours following a rain event. Samples were collected using existing dedicated facility autosamplers used for quality and compliance purposes. The autosampler was set up to collect time-based samples over 4-hr and 24-hr periods. For 24-hour samples each autosampler was set up to collect approximately 100 mL of sample in 15-minute increments over a 24-hour period and collection initiated on a Tuesday to eliminate effects from weekend flows. Samples were collected directly into a single, large 10-L glass container housed in a refrigerated cabinet. All samples were filtered on site through 90-mm glass fiber (GF/F) filters (Whatman, England) at the end of the collection period and collected into 1-L ultra-clean, pre-silanized, amber glass bottles containing 1 g sodium azide for preservation and 1 mL of 500-ppm ascorbic acid solution to quench any residual chlorine. Samples were then placed in a specialized hard-shell shipping case with ice packs and shipped via overnight courier to the SNWA analytical laboratory in Henderson, Nevada. All samples were extracted within 14 days of collection. Operational and water quality parameters measured by the utility at the time of sampling were gathered and compiled by CSM.

3.8.3 Blanks

For each facility, a travel blank was prepared at SNWA by filling a precleaned 1-L amber bottle with ultra-pure water. The travel blanks were sent to the facilities with the sampling bottles and sent back to SNWA without being opened by the facility.

3.8.4 Shipment and Storage

Once the samples were collected and prepared, they were shipped overnight to SNWA and CSM in ice-packed coolers. Upon arrival, samples were logged and stored at 4°C pending further analysis. For GC/MS-MS and LC/MS-MS analyses samples were extracted within two weeks. The extracts were stored at 4°C until analyses of extracts were completed. Field and laboratory blanks were processed like field samples.

Table 3-2. Sampling Locations at Full-Scale Facilities.

Facility	Type of receiving wastewater	Treatment train	Sampling location during occurrence monitoring	Sampling location during individual unit processes monitoring
Facility 1	Plant 1	Primary, secondary (nitrifying/denitrifying), tertiary filtration, disinfection (chlorine)	Raw sewage Final effluent	Primary effluent Secondary effluent Final effluent
	Plant 2	(a) Primary, secondary (nitrifying/ denitrifying), tertiary filtration, disinfection (chlorine), (b) primary, MBR	Raw sewage a Final effluent a	Primary effluent a Final effluent a MBR permeate b
Facility 2	Plant 3	Primary, secondary (partly nitrifying (a) and nitrifying/ denitrifying (b)), disinfection (chlorine)	Raw sewage Final effluent	Primary effluent a Primary effluent b Secondary effluent a Secondary effluent b
Facility 3	Plant 4	Receiving secondary effluent, Biological aerated filter (BAF), coagulation/flocculation, sedimentation, filtration, disinfection (chlorine)	Secondary effluent Final effluent	Secondary effluent BAF Filtered effluent
Facility 4	Plant 5	Primary, secondary (partly nitrifying/denitrifying), tertiary filtration, disinfection (UV)	Not sampled	Raw sewage Tertiary effluent Final effluent
Facility 5	Plant 6	Primary, secondary (partly nitrifying/denitrifying), tertiary filtration, disinfection (chloramination)	Raw sewage Final effluent	Tertiary effluent Final effluent
Facility 6	Plant 7	Primary, secondary (partly nitrifying/denitrifying), tertiary filtration, disinfection (chlorine)	Raw sewage Final effluent	Tertiary effluent Final effluent

Table 3-3. Overview of Selected Operational Information for Selected Full-Scale Study Sites.

Plant	Operational Capacity	Population Served	BOD Loading	BOD	HRT Primary	HRT Secondary	SRT	MLSS	NH_4-N Effluent	NO_3^--N Effluent	Disinfection Type
	Mgd		kg/day	mg/L	hours	hours	days	mg/L	mg/L	mg/L	
1	57.5	914K	18,130	207	1.65	2.7	15-48	3,418	3.9	3.8	NH_3/Cl_2
2[1]	10 25 gpm	150K	FS: 4,990 PP: 28	FS: NA PP: 165	FS: 3.2 PP: 3	FS: 3.2 PP: 6.7	FS: 19 PP: 10-15	FS: 3,400 PP: 7,950	FS: 1.2 PP: < 1 – 2.2	FS: 6.2 PP: 6.2	FS: NH_3/Cl_2 PP: NA
3[2]	N: 99.3 S: 34.1	1,300K	N: 103,154 S: 36,295	N: 216 S: 193	N: 1.6-2.1 S: 2.4-3.9	N: 6.7 S: 6.7	N: 6.1 S: 1.7	N: 3,004 S: 2,584	N: 3.3 S: 17.6	N: 8.52 S: 0.97	Cl_2
4	2.5-18	NA	NA	< 3 - 47	NA	NA	NA	NA	0.0 – 0.163	11.6	Cl_2
5	88.5	900K	52,550	149	1.5 – 2	3.4	6.17	3,006	0.27	15.02	UV
6	21	235K	17,327	224	NA	10.5 – 11.5	4-16	1,963	< 0.1 – 3.68	8.2	Hypochlorite
7	61.7	676K	12,086	59.3	2-4	2-4	10-15	2,610	< 0.1 – 0.84	20.5	Hypochlorite

Notes:

1 – Facility 2 comprised full-scale water reclamation plant and on-site membrane bioreactor pilot-plant

2 – Facility 3 comprised two parallel treatment trains.

BOD – biochemical oxygen demand

HRT – hydraulic residence time

SRT – solids retention time

MLSS – mixed liquor suspended solids

FS – after fine screen

PP – permeate

mgd – millions of gallons per day

gpm – gallons per minute

kg/day – kilograms per day

mg/L – milligrams per liter

NA – information not available or not applicable

CHAPTER 4.0

OCCURRENCE OF HPV HOUSEHOLD CHEMICALS IN SEWAGE AND TREATED WASTEWATER EFFLUENTS

4.1 Introduction

HPV chemicals released to the sewer exist as complex mixtures, which likely vary over time. Thus, the main objective of full-scale occurrence monitoring was to identify whether certain HPV household chemicals exhibit a consistent pattern of occurrence in raw sewage and treated wastewater effluents in treatment plants serving communities in different regions of the country. During field-scale monitoring, influent and effluent qualities of six different wastewater treatment facilities located in the western continental United States were screened for target organic chemicals using LC-MS/MS and GC-MS/MS as a sensitive and rapid screening tool for select HPV household chemicals. The wastewater facilities differed in treatment processes and operational parameters employed (e.g., BOD loading rate; solid retention time (SRT); hydraulic retention time (HRT), and biosolids treatment, (e.g., off-site versus on-site,), as well as wastewater composition (e.g., different proportions of residential and industrial discharge in the service area). Table 3-2 summarizes the treatment specifics and sampling locations of each plant selected for this study. More detailed information for each facility is provided in the Facility Survey Summary provided in Appendix A.

Different sampling strategies were conducted during full-scale occurrence monitoring. At each of six plants, 24-hour composite samples were collected from raw sewage and the final effluent. In addition, at plant 3, raw sewage composite samples were collected in 4-hour increments over a 24-hour period to assess daily variations. During a separate sampling campaign at plant 3, 24-hour composite samples of the raw influent were collected over an extended time period during the week and subsequent weekend: Thursday-Friday-Saturday-Sunday-Monday-Tuesday.

4.2 Occurrence of HPV Chemicals in Sewage and Treated Wastewater Effluents

4.2.1 Occurrence Based on 24-hr Composite Sampling at Six Full-Scale Facilities

Combined, average influent and effluent concentrations observed for all six plants are summarized in Table 4-1 and illustrated in Figures 4-1, 4-2, and 4-3 (Tier 1 and 2 and LAS compounds, respectively). Standard deviations are also provided in the table and figures. Due to apparent blank-contamination issues, results for bisphenol-A are not included in this analysis.

A review of influent and effluent analytical results for all facilities indicates a general pattern of occurrence and concentrations of HPV chemicals in raw sewage and treated wastewater. Usually 20 of the selected 26 target study compounds were present in the raw influent samples (Table 4-2). The herbicides atrazine and trifluralin and the pesticide acriflavine were generally not present in raw influent samples, whereas the herbicide simazine was detected at very low concentrations at plants 1, 2, and 7. The lack or low concentrations of these compounds in samples may reflect predominant outdoor applications, and subsequent release or run-off into storm sewers or surface waters rather than sanitary sewers. The majority of compounds present in personal care products generally appear in all influent samples with concentrations of 2-phenoxyethanol (a preservative with various uses) and menthol (a fragrance with various uses) consistently demonstrated the highest concentrations of all compounds. Most Tier 1 compounds exhibited concentration levels in raw influent samples generally in the microgram per liter (μg/L) range and at higher concentrations relative to Tier 2 compounds. The higher concentrations of Tier 1 compounds may be indicative of their classification as HPV chemicals manufactured or imported into the United States at greater volumes (> 1 million lb/yr) than Tier 2 compounds.

Table 4-1. Average HPV Compound Concentrations in Raw Sewage and Effluents (Plants 1, 2, 3, 5, 6, 7 combined). (Note: Concentrations reported are those determined in filtered samples)

Analyte	Analytical Method	Average Raw Influent[1] (ng/L)	Influent Standard Deviation (ng/L)	Average Final Effluent[2] (ng/L)	Effluent Standard Deviation (ng/L)
Tier 1 Compounds					
Atrazine	LC/MS-MS	1	2	1	2
Hexabromododecane	GC/MS-MS	<1,000	0	<100	0
Bisphenol-A	LC/MS-MS	n.r.	n.r.	n.r.	n.r.
Dibutyl phthalate	GC/MS-MS	2,850	1,930	590	410
Benzophenone	GC/MS-MS	1,50	480	220	200
Oxybenzone	LC/MS-MS	1,440	700	40	60
BHT	GC/MS-MS	385	280	70	50
2-Phenoxyethanol	GC/MS-MS	14,300	6,730	240	350
Vanillin	GC/MS-MS	3,211	3,120	265	160
Triclosan	LC/MS-MS	1,300	630	220	240
o-Phenylphenol	LC/MS-MS	1,800	1,850	75	110
	GC/MS-MS	1,800	1,700	80	115
Tier 2 Compounds					
Acriflavine	LC/MS-MS	<100	0	<10	0
Trifluralin	GC/MS-MS	<250	0	<25	0
Simazine	LC/MS-MS	4	7	5	8
3-indolebutyric acid	LC/MS-MS	870	620	170	115
DEET	LC/MS-MS	570	445	150	170
	GC/MS-MS	860	610	200	145
MGK-11	GC/MS-MS	<250	0	<25	0
Hydrocortisone	LC/MS-MS	270	80	6	15
Camphor	GC/MS-MS	1,650	309	13	35
Menthol	GC/MS-MS	10,300	6,800	115	250
2-methylresorcinol	GC/MS-MS	<1,000	0	<100	0
Propylparaben	LC/MS-MS	1,300	490	4	3
Isobutylparaben	LC/MS-MS	250	140	3	3
BHA	LC/MS-MS	160	60	70	105
	GC/MS-MS	190	60	90	130
Model Compounds					
Triclocarban	LC/MS-MS	215	160	70	42
LAS[3]	HPLC	5,980[3]	1.41	4.04[3]	0.18

Notes:

1 - raw influent values reflect an average of six values (n = 6)

2 - final effluent values reflect an average of six values (n = 6)

3 – LAS values reflect an average of two values (n = 2), values reported in micrograms-per-liter (µg/L)

ng/L – nanograms-per-liter; < value reflects the preliminary reporting limit; n.r. – not reported

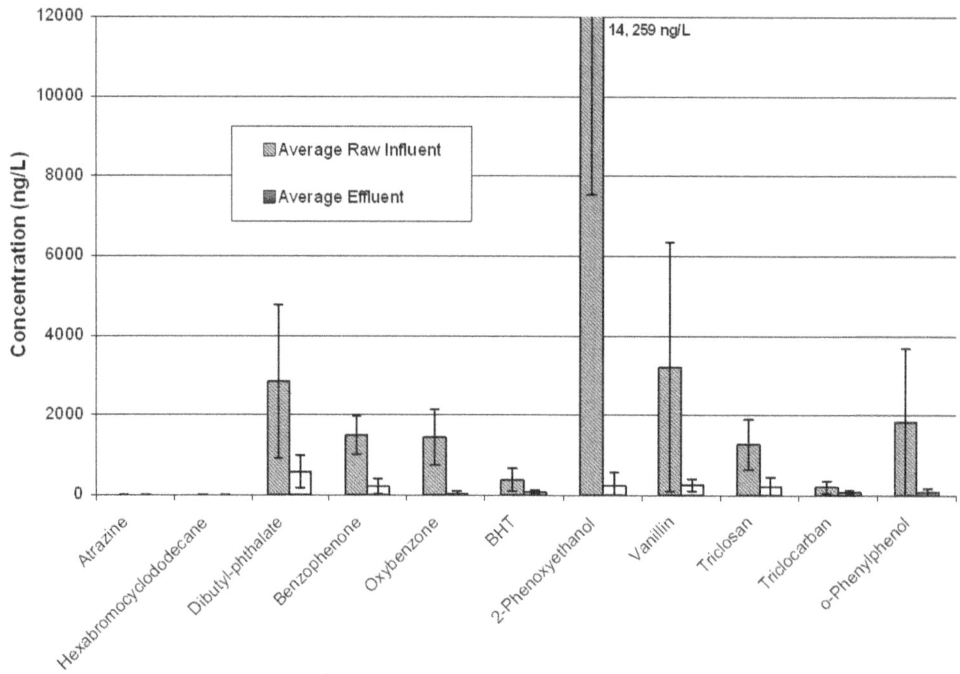

Figure 4-1. Occurrence of Tier 1 Compounds in Sewage and Treated Wastewater Effluents (average and standard deviation of six wastewater treatment plants) (Note: Concentrations reported are those determined in filtered samples).

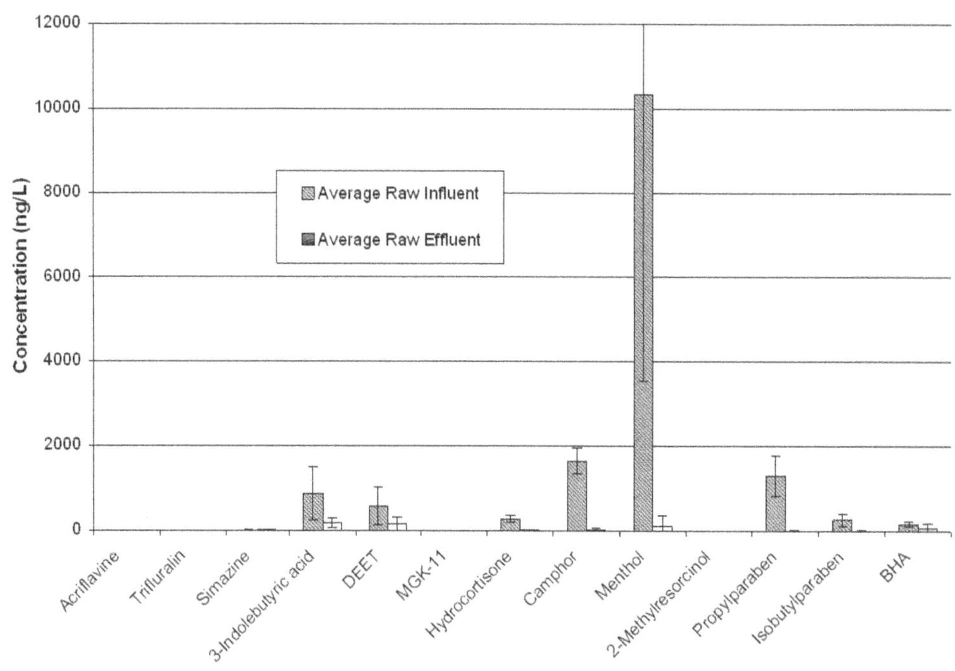

Figure 4-2. Occurrence of Tier 2 Compounds in Sewage and Treated Wastewater Effluents (average and standard deviation of six wastewater treatment plants) (Note: Concentrations reported are those determined in filtered samples).

In the wastewater effluent samples, the Tier 1 compounds dibutyl phthalate, benzophenone, 2-phenoxyethanol, vanillin, and triclosan exhibited average effluent concentrations exceeding 200 ng/L with dibutyl phthalate exhibiting the highest average concentration of approximately 600 ng/L. Only three out of 13 Tier 2 compounds, 3-indolebutyric acid, DEET, and menthol, exhibited concentrations exceeding 100 ng/L in wastewater treated effluents. Although influent and effluent concentrations exhibited significant variations (see standard deviation, Table 4-2), the good agreement between LC and GC methods for the analysis of o-phenylphenol, DEET, and BHA suggest that the observed variations were likely a function of variable consumption pattern and flow conditions rather than a result of analytical variability.

The concentrations of LAS compounds of different chain length measured in raw sewage and treated effluent are presented in Figure 4-3. The average concentration of LAS in raw sewage almost approaches 6 mg/L. This concentration is reduced in treated wastewater effluents to approximately 4,000 ng/L, representing the highest observed concentrations of all compounds targeted in this study.

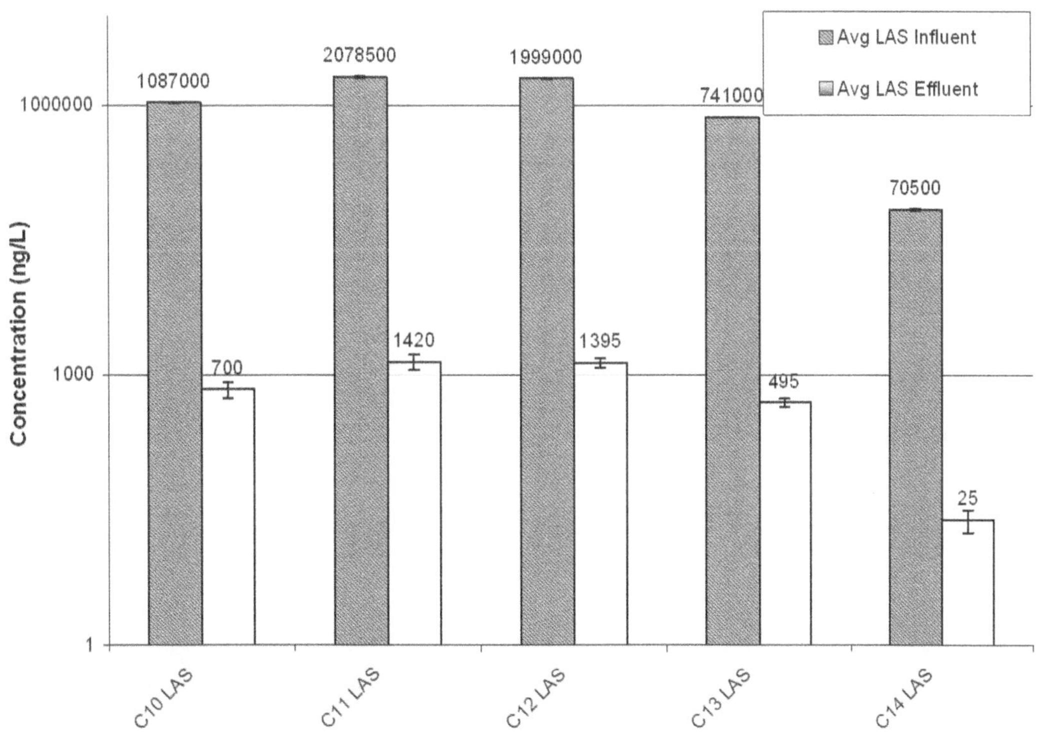

Figure 4-3. Occurrence of LAS in Sewage and Treated Wastewater Effluents (average and standard deviation of six wastewater treatment plants) (Note: Concentrations reported are those determined in filtered samples).

4.2.2 Daily and Weekly Occurrence Variations of HPV Household Chemicals

Four-hour composite samples were collected from the raw influent of one facility over a 24-hour period commencing at 8:00 am, Tuesday, October 12, 2005, until 8:00 am the following morning. Twenty-four-hr composite samples were also collected from the same facility over a six-day period (Thursday-Friday-Saturday-Sunday-Monday-Tuesday), commencing at 8:00 am each day, from Wednesday, October 20, until Tuesday, October 21, 2005. Samples were collected into individual glass containers (reflecting 4-hr composites) housed in a refrigerated cabinet. Operational and water quality parameters measured at the time of sampling are summarized in Table 4-2. These results confirm that the plant was operating within design specifications during time of sampling.

Analytical results for the 4-hr composite samples targeting HPV household chemicals are summarized in Table 4-3, and analytical results for the 24-hr composite samples are presented in Table 4-4. With the exception of atrazine, hexabromocyclododecane, acriflavine, trifluralin, MGK-11, and 2-methylresorcinol, all target study compounds were present in the raw influent of this facility. LAS was detected at significantly higher concentrations (on the order of μg/L) than the other target compounds. Concentrations were generally similar to those previously observed for this and other plants, with concentrations of 2-phenoxyethanol, dibutyl-phthalate, o-phenylphenol, and menthol generally being the highest (approaching low μg/L concentrations). Also, consistent with previous observations at other plants, most Tier 1 compounds were present in influent samples generally at higher concentrations relative to Tier 2 compounds.

Table 4-2. Measured Operational Parameters at Plant 3.
(Note: Influent concentrations reported are those determined in filtered samples)

Sample Time/Date	Flow[1] (mgd)	pH	Temp (°C)	Cond. (μS/cm)	NH_3-N (mg/L)	NO_3-N (mg/L)	PO_4-P (mg/L)	TOC (mg/L)
4-Hour Composite Samples								
Oct 12, 8:00-12:00	94	6.9	na	1004	34.4	0.03	7.11	40.5
Oct 12, 12:00-16:00	102	6.9	na	968	38.9	0.1	9.44	61.9
Oct 12, 16:00-20:00	87	6.9	na	935	24.5	0.09	10.3	69.5
Oct 12, 20:00-24:00	103	6.8	na	896	24.2	0	7.01	66.8
Oct 13, 00:00-4:00	76	6.7	na	912	26.5	0.04	7.23	63.9
Oct 13, 4:00-8:00	37	6.7	na	975	24	0.1	10.3	44.2
24-Hour Composite Samples								
Oct 20, 8:00 – Oct 21, 8:00	73	na	na	997	21.5	Na	8.07	77.2
Oct 21, 8:00 – Oct 22, 8:00	72	na	na	968	20.5	0.02	6.98	70.1
Oct 22, 8:00 – Oct 23, 8:00	71	na	na	956	24.4	0.02	5.99	66.3
Oct 23, 8:00 – Oct 24, 8:00	78	na	na	940	23.6	0	6.33	65.1
Oct 24, 8:00 – Oct 25, 8:00	75	na	na	914	16.8	0.14	6.66	77.5
Oct 25, 8:00 – Oct 26, 8:00	75	na	na	942	15	0	5.59	74.1

Notes: 1 – Average flow measured during sample event; na – not available.

Table 4-3. Target Compound Concentrations in Raw Sewage at Plant 3, 4-hr Influent Samples, Oct 12-13, 2005. (Note: Concentrations reported are those determined in filtered samples)

Analyte	Analytical Method	8:00-12:00	12:00-16:00	16:00-20:00	20:00-24:00	0:00-4:00	8:00-12:00	Avg. Conc.	Std. Dev.
				Tier 1 Compounds (ng/L)					
Atrazine	LC/MS-MS	<5.0	<5.0	<5.0	<5.0	<5.0	<5.0	<5.0	<5.0
Hexabromo-cyclododecane	GC/MS-MS	<500.0	<500.0	<500.0	<500.0	<500.0	<500.0	<500.0	<500.0
Dibutyl-phthalate	GC/MS-MS	2,255	1,970	2,270	2,365	2,235	2,365	2,243	145
Bisphenol-A	LC/MS-MS	111	159	332	172	168	161	184	76
Benzophenone	GC/MS-MS	675	1,330	1,310	1,125	770	580	965	331
Oxybenzone	LC/MS-MS	407	628	623	681	769	481	598	133
BHT	GC/MS-MS	255	520	565	585	550	315	465	142
2-Phenoxy-ethanol	GC/MS-MS	28,000	29,900	22,800	15,350	27,500	10,550	22,350	7,796
Vanillin	GC/MS-MS	3,910	4,265	5,210	4,545	3,930	4,210	4,345	485
Triclosan	LC/MS-MS	1,020	1,630	1,890	2,480	1,800	1,200	1,670	523
o-Phenylphenol	LC/MS-MS	3,790	2,690	2,900	2,560	2,020	3,370	2,888	624
	GC/MS-MS	5,500	4,100	4,400	3,400	2,465	5,300	4,194	1,159
				Tier 2 Compounds (ng/L)					
Acriflavine	LC/MS-MS	<5.0	<5.0	<5.0	<5.0	<5.0	<5.0	<5.0	<5.0
Trifluralin	GC/MS-MS	<125	<125	<125	<125	<125	<125	<125	<125
Simazine	LC/MS-MS	5.8	<5.0	5.1	5.8	7.0	6.2	5	3
3-Indolebutyric acid	LC/MS-MS	311	491	343	287	289	383	351	78
DEET	LC/MS-MS	199	297	412	474	469	256	351	116
	GC/MS-MS	520	790	995	1,155	1,040	610	852	253
MGK-11	GC/MS-MS	<125	<125	<125	<125	<125	<125	<125	<125
Hydrocortisone	LC/MS-MS	174	246	210	203	184	139	193	36
Camphor	GC/MS-MS	765	1,305	1,700	2,075	1,565	1,235	1,441	448
Menthol	GC/MS-MS	13,900	14,650	10,900	10,700	13,050	12,000	12,533	1,608
2-Methyl-resorcinol	GC/MS-MS	<500	<500	<500	<500	<500	<500	<500	<500
Propylparaben	LC/MS-MS	1,060	1,640	1,390	1,430	1,530	917	1,328	280
Isobutylparaben	LC/MS-MS	258	383	319	330	379	230	317	62
BHA	LC/MS-MS	202	163	150	140	157	137	158	24
	GC/MS-MS	220	200	175	165	160	160	180	25
				Model Compounds[1]					
Triclocarban	LC/MS-MS	242	485	360	453	457	217	369	116
LAS (μg/L)	LC-MS	1,859	4,336	5,177	5,931	4,861	3,475	4,273	1,441

Notes:

ng/L – nanograms per liter

< value reflects reporting limit

NA – not analyzed

1 – LAS concentrations reported in micrograms per liter (μg/L)

Table 4-4. Target Compound Concentrations in Raw Sewage at Plant 3, 24-hr Influent Samples, Oct. 20-26, 2005. (Note: Concentrations reported are those determined in filtered samples)

Analyte	Analytical Method	10/20-10/21	10/21-10/22	10/22-10/23	10/23-10/24	10/24-10/25	10/25-10/26	Avg. Conc.	Std. Dev.
Tier 1 Compounds (ng/L)									
Atrazine	LC/MS-MS	<5.0	<5.0	<5.0	<5.0	<5.0	<5.0	<5.0	<5.0
Hexabromo-cyclododecane	GC/MS-MS	<500.0	<500.0	<500.0	<500.0	<500.0	<500.0	<500.0	<500.0
Dibutyl-phthalate	GC/MS-MS	3,250	2,650	2,145	6,200	7,100	3,100	4,074	2,052
Bisphenol-A	LC/MS-MS	515	666	574	489	519	771	589	109
Benzophenone	GC/MS-MS	1,275	1,005	890	960	2,075	1,165	1,228	438
Oxybenzone	LC/MS-MS	666	695	853	926	636	636	735	124
BHT	GC/MS-MS	240	380	465	645	475	660	478	160
2-Phenoxy-ethanol	GC/MS-MS	23,650	23,550	20,050	21,050	21,850	23,400	22,258	1,511
Vanillin	GC/MS-MS	3,295	3,280	2,440	2,800	3,065	2,915	2,966	324
Triclosan	LC/MS-MS	1,650	1,590	1,960	1,860	1,320	2,120	1,750	288
o-Phenylphenol	LC/MS-MS	5,140	5,870	1,940	734	3,720	5,130	3,756	2,035
	GC/MS-MS	4,700	5,850	1,645	895	3,250	4,300	3,440	1,891
Tier 2 Compounds (ng/L)									
Acriflavine	LC/MS-MS	<5.0	<5.0	<5.0	<5.0	<5.0	<5.0	<5.0	<5.0
Trifluralin	GC/MS-MS	<125	<125	<125	<125	<125	<125	<125	<125
Simazine	LC/MS-MS	<5.0	<5.0	<5.0	<5.0	<5.0	<5.0	<5.0	<5.0
3-Indolebutyric acid	LC/MS-MS	1,020	839	627	609	526	484	684	205
DEET	LC/MS-MS	425	340	378	479	375	304	384	62
	GC/MS-MS	855	770	905	1,185	900	760	896	155
MGK-11	GC/MS-MS	<125	<125	<125	<125	<125	<125	<125	<125
Hydrocortisone	LC/MS-MS	206	226	213	228	195	188	209	16
Camphor	GC/MS-MS	1,650	1,535	1,670	2,085	1,415	1,470	1,638	241
Menthol	GC/MS-MS	10,200	9,150	10,600	10,550	8,650	10,100	9,875	795
2-Methyl-resorcinol	GC/MS-MS	<500.0	<500.0	<500.0	<500.0	<500.0	<500.0	<500.0	<500.0
Propylparaben	LC/MS-MS	1,430	1,270	1,400	1,450	1,280	1,400	1,372	77
Isobutylparaben	LC/MS-MS	271	274	310	307	266	272	283	20
BHA	LC/MS-MS	246	341	267	254	259	242	268	37
	GC/MS-MS	190	315	225	220	195	195	223	47
Model Compounds (ng/L)[1]									
Triclocarban	LC/MS-MS	337	284	368	436	370	340	356	50
LAS (μg/L)	LC-MS	3,590	4,002	5,072	5,922	4,058	4,253	4,483	858

Notes:

ng/L – nanograms per liter

< value reflects reporting limit

1 – LAS concentrations reported in micrograms per liter (μg/L)

Daily mass loading variations of Tier 1 target compounds are presented in Figure 4-4. Only aqueous concentrations of target compounds were considered in this effort. Therefore, mass loading in the context of this study refers to mass dissolved in wastewater and does not include mass that is associated with particles or biosolids.

The facility monitored exhibited a typical diurnal flow pattern. With the exception of o-phenylphenol and triclosan, the mass loading of all other Tier 1 compounds follows the same trend suggesting a rather constant discharge of these chemicals into the sewer and a release of household chemicals that is directly coupled to water consumption. Hourly observed concentrations and mass loading values at this plant indicate higher usage for most compounds during the morning hours (corresponding to 12:00-16:00 samples), such as the widely used germicide, o-phenylphenol, exhibited the highest loading between 8.00 am and 12 pm. However, triclosan, an antimicrobial used in many cleaning agents, exhibited the highest loading between 8 pm and midnight. This pattern might be a consequence of cleaning activities at the end of the day.

Loadings of Tier 2 compounds are significantly lower than Tier 1 compounds (data not shown for Tier 2 compounds), but they also follow the general diurnal trend with the exception of DEET, the key ingredient of mosquito repellents. The highest loading of DEET was observed between 8 pm and midnight. Because DEET is applied to the skin, it can easily be washed off while taking a shower.

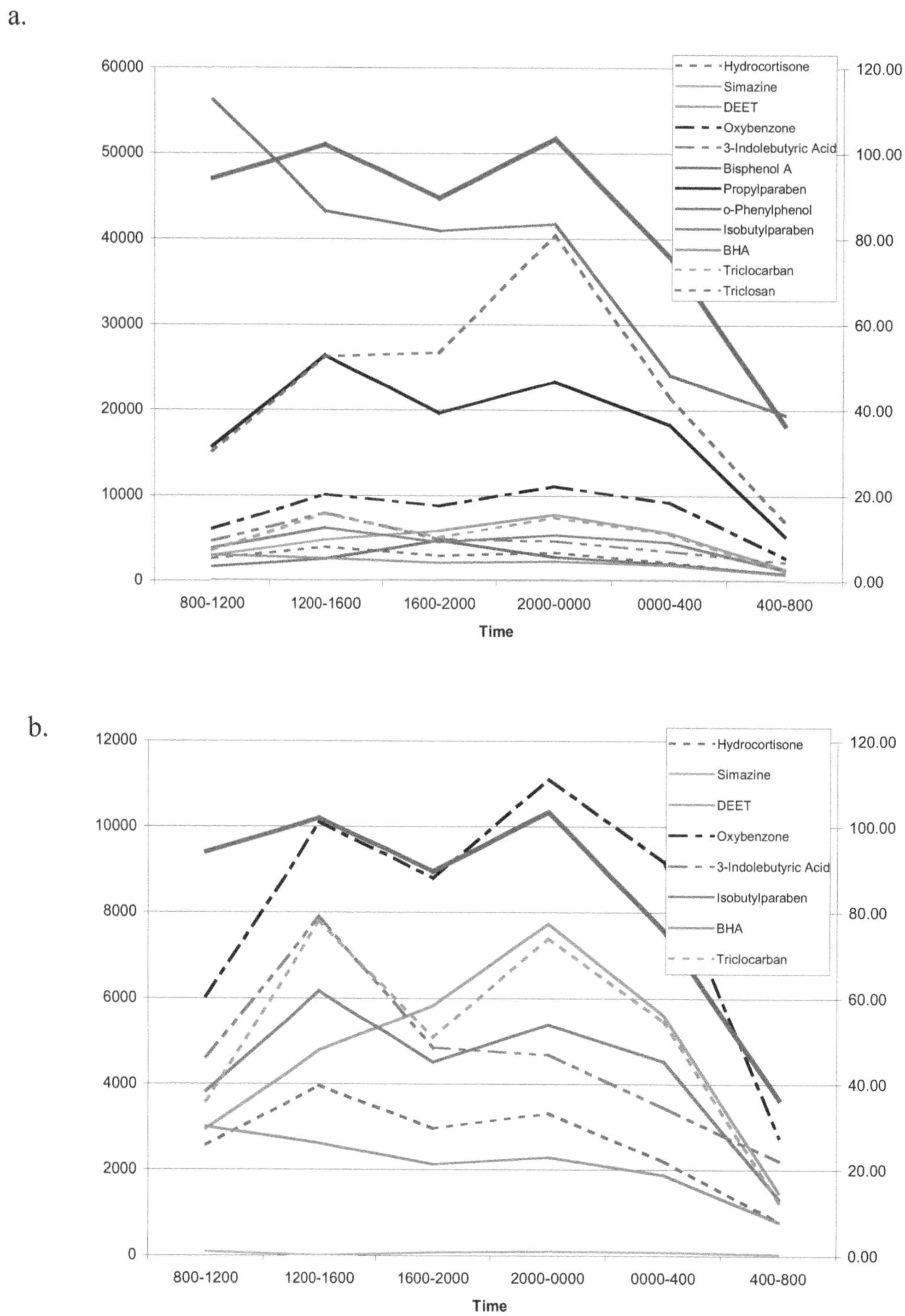

Figure 4-4. a) Daily Mass Loading Variations and b) Daily Mass Loadings Less than 12,000 (g/day) for Tier 1 Household Chemicals at Plant 3. (Note: Concentrations considered are those determined in filtered samples)

To evaluate the weekly occurrence variation of the target compounds, mass loading values for concentrations detected in the 6-day, 24-hr composite samples were calculated using the average influent flow (Table 4-2) measured during each of the sampling events. The aqueous phase mass loading values for detected Tier 1 and 2 compounds are presented in Figure 4-5 and 4-6, respectively. Higher weekend loading values for some compounds (i.e., dibutyl phthalate, oxybenzone, triclocarban, camphor, and DEET) may further indicate that many of these compounds have significant usage in household products. Conversely, lower weekend loading values observed for benzophenone and o-phenylphenol may indicate that these compounds also have significant uses other than household (i.e., industrial).

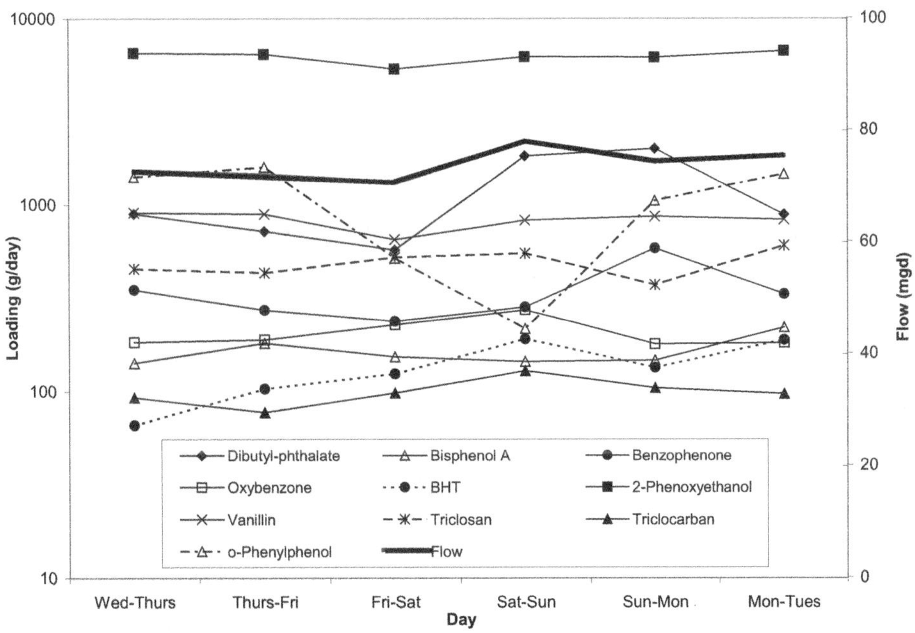

Figure 4-5. Six-Day Loading of Tier 1 Household Chemicals at Plant 3. (Note: Concentrations considered are those determined in filtered samples)

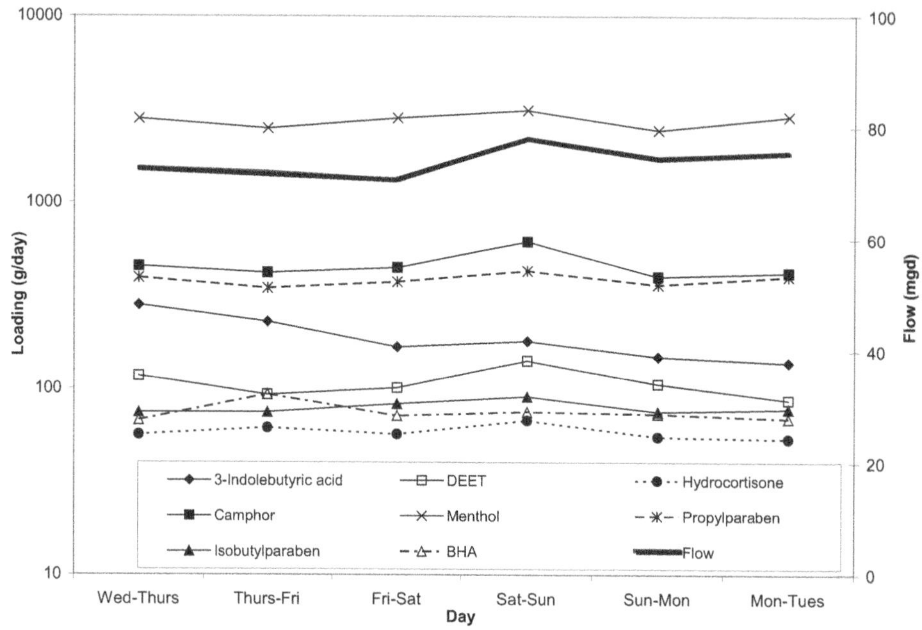

Figure 4-6. Six-Day Loading of Tier 2 Household Chemicals at Plant 3.
(Note: Concentrations considered are those determined in filtered samples)

Total daily and weekly linear alkylbenzene sulfonate mass loading for this facility was evaluated by averaging LAS aqueous phase concentrations over hourly or daily flow rates. Hourly and weekly LAS mass loadings for this facility are illustrated in Figures 4-7 and 4-8, respectively.

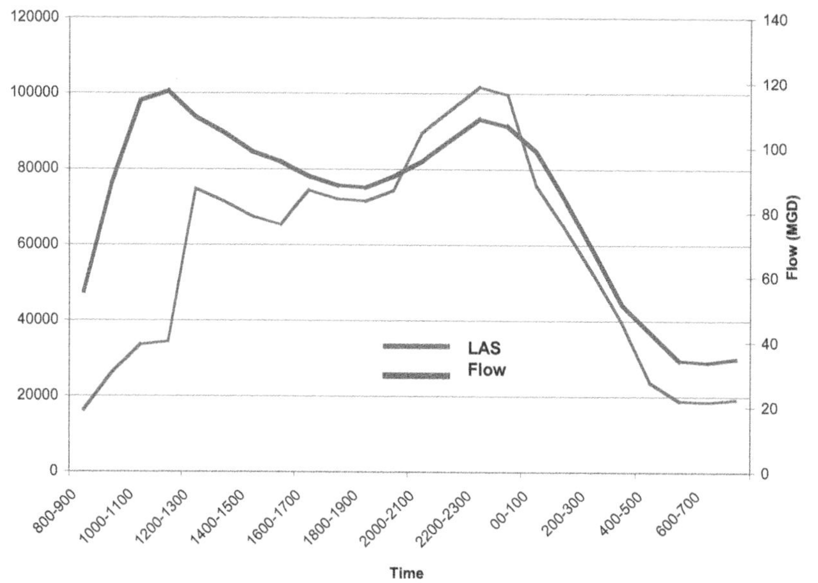

Figure 4-7. Daily LAS Mass Loading Entering at Plant 3.
(Note: Concentrations considered are those determined in filtered samples)

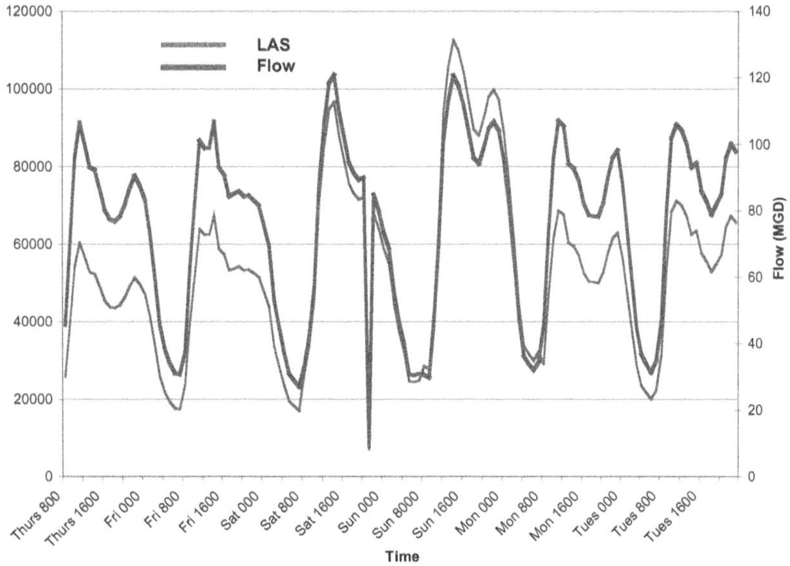

Figure 4-8. Weekly LAS Mass Loading Entering at Plant 3.
(Note: Concentrations considered are those determined in filtered samples)

Similar to previous observations with Tier 1 and 2 compounds, LAS loading exhibits a pattern following the diurnal flow pattern of this facility. This finding is expected since LAS are widely used as surfactants in many cleaning products and its use seems to be closely coupled with human activities and fluctuations in water consumption. The diurnal loading pattern of LAS was confirmed by collecting daily composite samples on seven consecutive days (Figure 4-9). Weekly sampling, however, revealed a significant higher loading of LAS during the weekend, which might be a consequence of more cleaning/laundry activities during the weekend.

4.2.2 Overall Treatment Efficiency

For Tier 1 and 2 compounds detected in raw sewage, the removal efficiency of full-scale conventional wastewater treatment plants (determined by comparing final effluent concentrations to raw influent concentrations as determined in filtered samples) was found to be variable, but generally exceeded 80% (Figure 4-9). 2-phenoxyethanol, hydrocortisone, camphor, propylparaben, and isobutylparaben were consistently removed by 98% or higher. Removal for BHT, BHA, DEET, 3-indolebutyric acid, and triclocarban was 69%, 64%, 69%, 65%, and 64%, respectively. Removal of these compounds also exhibited the highest variability. Effluent concentrations for atrazine and simazine were less than 5 ng/L, and, given their low occurrence levels, the removal percentages for these two compounds were not quantifiable.

The removal of Tier 1 and Tier 2 compounds occurring at significant levels was independent from solids retention times (SRT) varying from 1.7 to 32 days (Figure 4-10). With the exception of menthol, 2-phenoxyethanol, vanillin, and propylparaben did not exhibit an improved removal with longer SRTs. Removal of dibutyl phthalate varied between 65% and 95% with increasing SRT. An increasing mixed liquor aeration time or hydraulic retention time (HRT) varying from 2 to 6 hours did not result in an improved removal of target compounds (Figure 4-11).

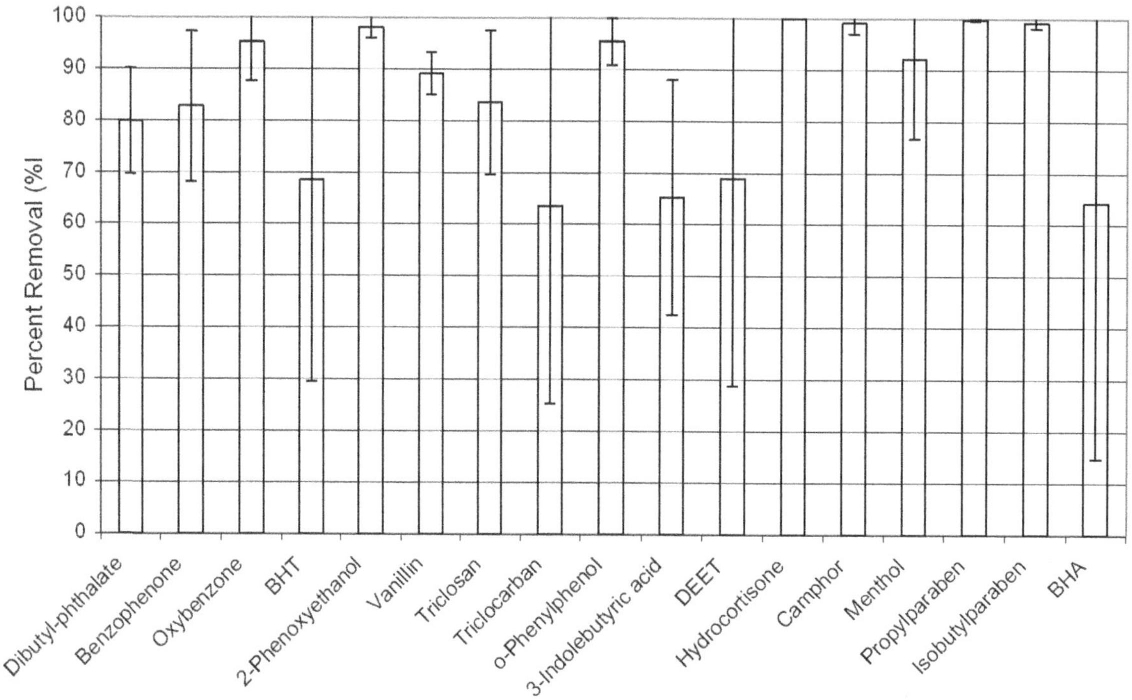

Figure 4-9. Average Removal of Select Tier 1 and Tier 2 Compounds during Wastewater Treatment. (Note: Concentrations considered are those determined in filtered samples)

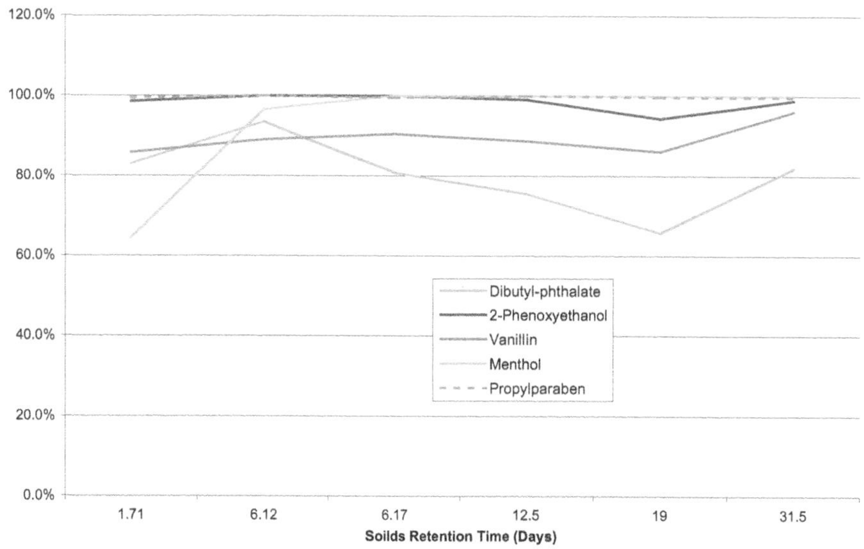

Figure 4-10. Percent Removal of Select Tier 1 and Tier 2 Compounds with Increasing Solids Retention Time. (Note: Concentrations considered are those determined in filtered samples)

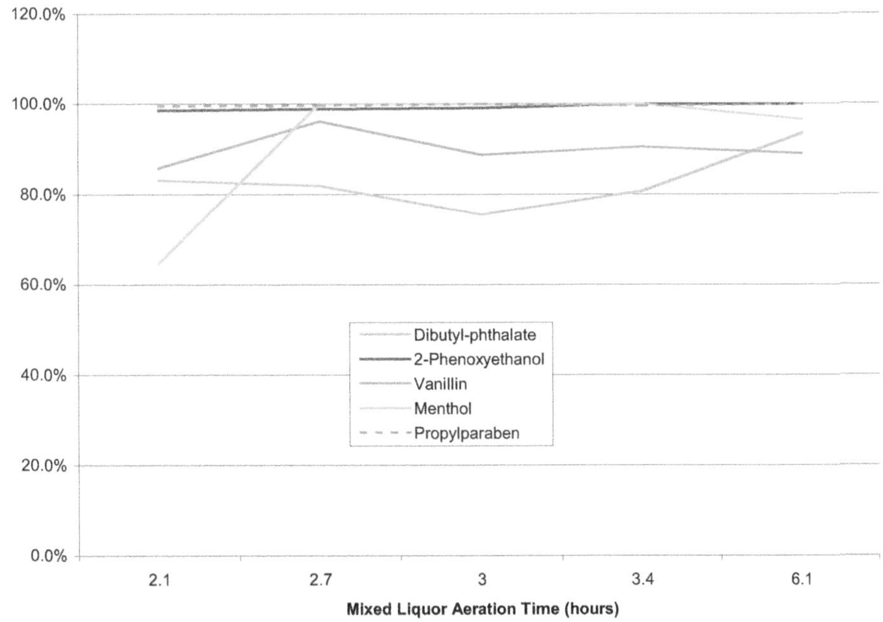

Figure 4-11. Percent Removal of Select Tier 1 and Tier 2 Compounds with Increasing Hydraulic Retention Time. (Note: Concentrations considered are those determined in filtered samples)

4.3 Conclusions

Of the 26 household chemicals targeted in this study, 20 compounds were consistently detected in glass fiber filtered raw influents of full-scale wastewater treatment plants. Chemicals that are primarily used in products applied outdoors, such as pesticides (e.g., simazine, atrazine, trifluralin, and acriflavine), or very hydrophobic compounds, such as the flame retardant hexabromocyclododecane, were generally not present in raw influent samples. The majority of compounds present in personal care and cleaning products generally appeared in all influent samples with concentrations of 2-phenoxyethanol (a preservative with various uses) and menthol (a fragrance with various uses) consistently exhibiting the highest concentrations of all compounds. In general, aqueous concentrations of Tier 1 compounds in raw sewage were in the microgram-per-liter range, whereas Tier 2 compounds were present at several hundred nanograms-per-liter. In treated wastewater effluents, the Tier 1 compounds dibutyl phthalate, benzophenone, 2-phenoxytethanol, vanillin, and triclosan exhibited average effluent concentrations exceeding 200 ng/L with dibutyl phthalate exhibiting the highest average concentration of approximately 600 ng/L. Only three out of 13 Tier 2 compounds, 3-indolebutyric acid, DEET, and menthol, exhibited aqueous concentrations exceeding 100 ng/L in filtered wastewater treated effluents. The average concentration of LAS in filtered raw sewage almost approached 6 mg/L and the concentration was reduced in wastewater treated effluents to approximately 4,000 ng/L, representing the highest observed concentrations of all compounds targeted in this study. Daily mass loading variations of all Tier 1 compounds as determined in the aqueous phase suggested a rather constant discharge of household chemicals to sewage with discharge directly coupled to water consumption. Daily mass loadings of Tier 2 compounds were significantly lower than Tier 1 compounds but also followed the general diurnal trend in occurrence. Higher weekend loading values for some compounds (i.e., dibutyl phthalate, oxybenzone, triclocarban, camphor, and DEET) may further indicate that many of these compounds have significant usage in household products. Conversely, lower weekend loading values observed for benzophenone and o-phenylphenol may indicate that those compounds also have significant uses other than household (i.e., industrial). Daily LAS mass loading as determined in the aqueous phase exhibited an occurrence pattern following the common diurnal flow pattern. This finding is expected since LAS is widely used as surfactants in many cleaning products and its use seems to be closely coupled with human activities and water consumption fluctuations. Weekly sampling, however, revealed a significant higher loading of LAS during the weekend, which might be a consequence of more cleaning/laundry activities during the weekend

Occurrence in the aqueous phase of sewage seems to be not only dependent upon production volume but also dependent upon compound properties (Figure 4-12). Of the three compounds produced between 10 and 50 million pound per year (BHT, dibutyl phthalate, and 2-phenoxyethanol), 2-phenoxyethanol is the most water soluble (log K_{ow} of 1.16) exhibiting the highest mass loading in raw sewage. BHT is the most hydrophobic compound (log K_{ow} of 5.1) and exhibiting the lowest aqueous mass loading. A similar trend was observed for compounds produced between 1 and 10 million lb/yr with vanillin, being the most hydrophilic (log K_{ow} 1.21), exhibiting the highest mass loading and triclocarban, being the most hydrophobic (log K_{ow} 4.9), exhibiting the lowest aqueous phase mass loading. Compounds produced in smaller volume bins have log K_{ow} values ranging from 2.38-3.79 and for these compounds no clear relationship between production volume and hydrophobicity was obvious.

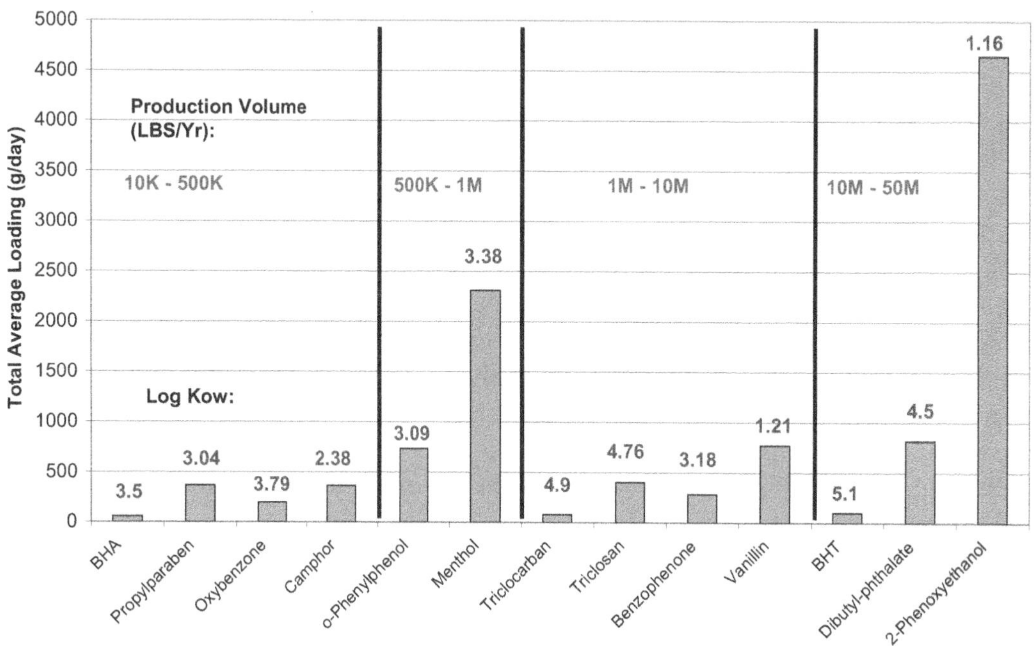

Figure 4-12. Total Average Loading of HPV Household Chemicals and Production Volume. (Note: Concentrations considered are those determined in filtered samples)

CHAPTER 5.0

FATE OF HPV HOUSEHOLD CHEMICALS DURING CONVENTIONAL AND ADVANCED WASTEWATER TREATMENT UNIT OPERATIONS

5.1 Introduction

Removal and destruction of household chemicals were further evaluated for individual unit processes using laboratory-, pilot-, and/or full-scale processes. The following treatment processes were evaluated: activated-sludge, UV, chlorine, and chloramine disinfection, MBR, ozonation, and advanced oxidation (ozone-H_2O_2). Activated-sludge, ozone, and advanced oxidation treatments were evaluated at the laboratory scale using flow-through systems. MBR was evaluated at the pilot scale. Activated sludge and UV, chlorine, and chloramine disinfections were evaluated at full scale. Time-based 4-hr composite samples were collected for full- and pilot-scale unit operations. This approach considered the retention time of the individual unit to ensure synoptic sampling and collection of hydraulically corresponding samples through a particular treatment process. Table 3-2 summarizes the treatment specifics and full- and pilot-scale sampling locations for individual unit processes of each plant selected for this study. More detailed information for each facility is provided in the Facility Survey Summary provided in Appendix A. The participating utilities assisted with sampling logistics and provided operational information, such as flow, BOD loading, SRT, HRT, and data for the time of sampling (e.g., pH, total suspended solids, mixed liquor total suspended solids, TOC, ammonia, and nitrate).

5.2 Removal of HPV Household Chemicals during Wastewater Treatment Processes

5.2.1 Biological Treatment: Activated Sludge and MBR

Facility 1 operates a full-scale activated-sludge plant (Plant 1) and a pilot-scale MBR plant (Plant 2b) where both receive sewage from the same service area. The service area has no combined sewers; therefore, no stormwater run-off is treated. The full-scale activated-sludge process (~60 mgd) uses both nitrification and denitrification. The membrane bioreactor was fed with domestic primary effluent (40 gpm). The MBR consists of two anoxic tanks, one aeration tank and one membrane tank in series with a total working capacity of approximately 11,700 gallons. Each anoxic tank is equipped with a mechanical mixer with an approximate working volume of 1,670 gallons. The primary effluent is fed to the first anoxic tank through a 3-inch feed spool. The water level in the first anoxic tank is monitored to control the feed. The first anoxic tank also recovers mixed liquor from the aeration tank. The aeration tank has a working volume of approximately 6,700 gallons, which uses process air via 38 10-inch fine bubble diffusers. A 5-hp recirculation pump in the aeration tank pumps mixed liquor to both the membrane tank (120 gpm) and the first anoxic tank (120 gpm). Sludge is wasted from the aeration tank. The membrane tank has an aerobic operating volume of 1,588 gallons: Two membrane cassettes, each containing 10 Zenon hollow fiber membrane elements, are situated in the membrane tank. The total membrane area is 4,730 ft^2. The nominal membrane pore size is 0.04 μm. The membrane tank includes a recycle stream to the aeration tank at 80 gpm.

Tier 1 and 2 concentrations of the influent and effluent of activated-sludge and MBR treatments are summarized in Table 5-1. Samples were collected from the primary treated effluent of each plant, while secondary treated effluent was collected at the secondary clarifier of Plant 1 and from the permeate of the MBR pilot plant. The removal efficiencies of the Tier 1 and 2 target compounds for the two processes are illustrated in Figure 5-1. In general, the two processes appear to have a similar capacity to remove the target compounds with most compounds experiencing greater than 80% removal for both processes. The MBR process did not appear to be effective in removing bisphenol-A and DEET, while the activated sludge process was not as efficient in removing BHT, 3-indolebutyric acid, hydrocortisone, and BHA.

Table 5-1. Comparison of Activated-Sludge and MBR Biological Treatment Processes. (Note: Concentrations reported are those determined in filtered samples)

Analyte[1]	Plant 1 Primary Effluent	Plant 1 Secondary Effluent	Plant 2b MBR Influent	Plant 2b MBR Effluent
Tier 1 Compounds (ng/L)				
Atrazine	<100	<5	<10	<10
Hexabromocyclododecane	<100	<100	<100	<100
Dibutyl phthalate	3,280	530	1,960	387
Bisphenol-A	138	<50	241	225
Benzophenone	1,310	109	3,110	60
Oxybenzone	350	<5	47	<10
BHT	299	53	103	<25
2-Phenoxyethanol	11,440	226	15,160	173
Vanillin	8,440	560	5,150	410
Triclosan	1,300	83	480	47
o-Phenylphenol	730	35	1,160	17
Tier 2 Compounds (ng/L)				
Acriflavine	<100	<5	<10	<10
Trifluralin	<25	<25	<25	<25
Simazine	<100	13	<10	<10
3-Indolebutyric acid	1,400	513	2200	53
DEET	156	25	74	44
MGK-11	<25	<25	<25	<25
Hydrocortisone	135	17	58	<10
Camphor	1,340	<50	312	<50
Menthol	8,440	<50	11,760	<50
2-Methylresorcinol	<100	<100	<100	<100
Propylparaben	2,020	1.4	1,480	0.88
Isobutylparaben	400	<1.25	333	0.29
BHA	153	24	128	2.7
Model Compounds (ng/L)				
Triclocarban	284	108	29	73
LAS[3]	na	na	na	na

Notes: 1 – All values reported were analyzed by LC/MS-MS unless otherwise noted
"<" – value reflects the preliminary reporting limit for that compound.
na – not analyzed

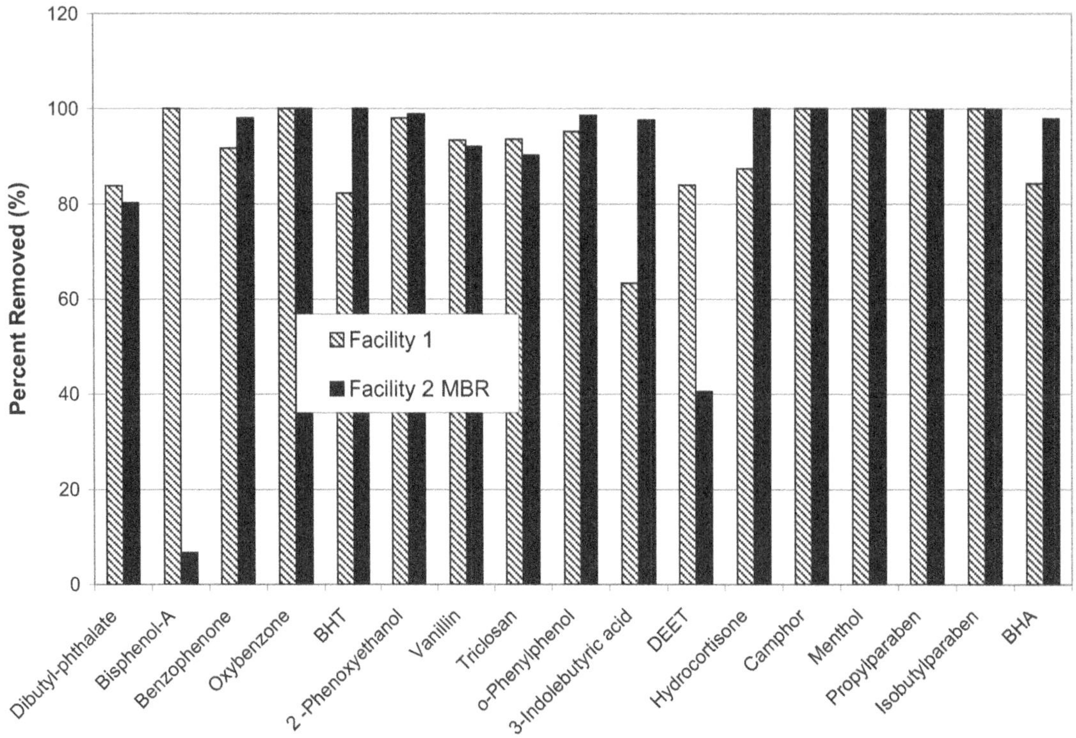

Figure 5-1. Comparison of Activated-Sludge and MBR Biological Treatment Processes.

5.2.1.1 Laboratory-Scale Activated Sludge Treatment

The removal of select target indicator, Tier 1, and Tier 2 compounds by activated sludge (AS) treatment was evaluated using a laboratory-scale continuous-stirred tank reactor. This lab-scale system consists of an aerobic basin (aerated) followed by a clarifier where sludge is recycled from the clarifier to the aerobic basin. Removals were determined under different SRT conditions.

The laboratory-scale AS system was seeded with biosolids from Plant 3's fully nitrified/denitrified aerobic basin and acclimated with a synthetic feed solution for 2 months. Table 5-3 summarizes the operational conditions of the AS system during start-up. The feed solution contained dextrose and glutamic acid as the carbon source, ammonia, and other important macronutrients and micronutrients. The feed solution was buffered to pH~8 with sodium bicarbonate. The system was acclimated when the effluent ammonia and basin TSS concentrations were stable. After acclimation, the trace organic compounds were added to the feed solution and continuously fed to the system. Each compound had an initial concentration around 500 μg/L in the feed.

Five experiments were performed representing differing SRT conditions, as shown in Table 5-2. Table 5-3 shows the list of compounds and their properties that were evaluated in the experiments. During all the experiments, the pH was maintained between 8.0 and 8.6, temperature was maintained between 14°C and 21°C and DO was maintained above 2.0 mg/L. In addition, for all experiments TOC and ammonia removals, either partially or completely, and the production of nitrate was observed supporting the presence of nitrification conditions.

Table 5-2. Operational Conditions of the AS system during Start-Up and Experiments 1-5.

Parameter	Units	Start-up & Exp I	Exp II	Exp III	Exp IV	Exp V
Volume of sludge basin	L	14	14	14	14	14
Volume of clarifier	L	8	8	8	8	8
Sludge recycle rate	L/d	117	117	117	117	117
Influent flow rate	L/d	35	50	28	28	27
Flow rate ratio	-	3.3	2.3	4.1	4.1	4.3
Waste flow rate	L/d	0	0	9.8	2.6	1.06
TSS aeration basin	mg/L	3510 (±612)	4280 (±407)	320 (±63)	1450 (±227)	2090 (±306)
TSS effluent/waste	mg/L	14 (±3)	37 (±16)	21 (±6)	28 (±9)	26 (±6)
HRT	hours	18	14	18	18	18
SRT	days	120 (±27)	35 (±10)	1.2 (±0.1)	4.7 (±0.2)	10.1 (±0.8)

Table 5-3. Properties for Organic Compounds used in Lab-Scale Activated Sludge Experiments.

Name	CAS #	Mol Wt (g/mol)	Henry's Constant (m3-atm/mol)	Water Solubility (mg/L)	pKa	Pre-dominant Ionization (pH 8.5)	Log D (pH 8.5)	Application
Butylated Hydroxyanisole	25013-16-5	180.3	1.17E-06	213	11.19 (acid)	Uncharged	3.5	Personal care product
Carbamazepine	298-46-4	236.3	1.08E-10	17.7	0.37 (base) -3.55 (base)	Uncharged	2.67	Pharmaceutical
DEET	134-62-3	191.3	2.08E-08	912	-	Uncharged	1.96	Personal care product
Diclofenac	15307-86-5	296.2	4.73E-12	2.37	4.15 (acid) -2.18 (base)	Negative	0.28	Pharmaceutical
Dilantin	57-41-0	252.3	1.02E-11	32	9.13 (acid) 19.83 (acid)	Uncharged & Negative	2.19	Pharmaceutical
Fenofibrate	49562-28-9	360.8	4.46E-09	0.418	-	Uncharged	5.19	Pharmaceutical
Gemfibrozil	25812-30-0	250.3	1.19E-08	10.9	4.75 (acid)	Negative	0.79	Pharmaceutical
Ibuprofen	79261-49-7	206.3	1.50E-07	21	4.91 (acid)	Negative	0.47	Pharmaceutical
Indolebutyric acid	133-32-4	203.2	1.28E-11	250	4.83 (acid) 0.4 (base)	Negative	-0.79	Household Chemical
Naproxen	22204-53-1	230.3	3.39E-10	159	4.15 (acid)	Negative	-0.6	Pharmaceutical
Phenacetine	62-44-2	179.2	2.13E.10	766	14.13 (acid)	Uncharged	1.63	Pharmaceutical
Phenylphenol	90-43-7	170.2	1.05E-06	700	9.99 (acid)	Uncharged	2.94	Personal care product
Primidone	125-33-7	218.3	1.94E-10	500	11.3 (acid) 12.25 (acid)	Uncharged	0.4	Pharmaceutical
Propylparaben	94-13-3	180.2	6.37E-09	500	8.23 (acid)	Uncharged & Negative	2.42	Personal care product
Propyphenazone	479-92-5	230.3	1.84E-09	3.00E+06	0.11 (base) -4.92 (base)	Uncharged	1.94	Pharmaceutical
Sulfamethoxazole	723-46-6	253.4	6.93E-08	610	6.16 (acid) 1.97 (base) 0.24 (base)	Negative	-0.07	Pharmaceutical
Triclosan	3380-34-5	289.5	4.99E-09	10	7.8 (acid)	Negative	4.11	Personal care product
Trimethoprim	738-70-5	290.3	2.39E-14	400	7.16 (base) -0.9 (base)	Uncharged	1.00	Pharmaceutical

Table 5-4 summarizes the removal of the studied organic compounds for SRT values between 1 to 120 days. Primidone, carbamazepine, and diclofenac were poorly removed at SRTs exceeding 120 days. Clara et al. (2005) also observed poor removal of carbamazepine during activated sludge treatment. Gemfibrozil, dilantin, and trimethoprim were poorly removed for SRT < 10 days, however, these compounds can be potentially better removed at higher SRT values. Naproxen and ibuprofen were well removed for SRTs exceeding 1 day after an acclimation period. During acclimation the biosolids adapt to degrading a particular compound. Results presented in Figure 5-2 illustrate the acclimation period for naproxen and ibuprofen (SRT = 1 d) where the effluent concentrations decrease versus time until steady-state conditions are reached. These polar compounds (negatively charged) were not shown to adsorb to biosolids in batch experiments (Section 5.2.1.2), thus biodegradation is the dominant removal mechanism for these compounds. There is also an observed acclimation period for phenacetine and triclosan

(SRT = 1d) indicating the presence of biodegradation, but sorption is also important based on batch sorption tests (Section 5.2.1.2) with phenacetine and Kinney et al. (2006) showed sorption is the dominant removal mechanism for triclosan. Like triclosan (log D = 4.11 at 8.5), fenofibrate has a high log D of 5.19 at 8.5, which indicates removal most likely due to sorption. Also, sulfamethoxazole, DEET, indolebutyric acid, phenylphenol, and propylparaben were well removed for SRTs exceeding 5 days after an acclimation period (Figure 5-2). Note, sulfamethoxazole and DEET were not removed at a SRT of 1 day (indolebutyric acid, phenylphenol, and propylparaben was not studied at SRT = 1d). The presence of an acclimation period and low sorption potential (log D -0.79, -0.04) indicates biodegradation as the likely dominant removal mechanism for sulfamethoxazole, DEET. Indolebutyric acid, phenylphenol, and propylparaben have low D values of 1.96, 2.94, and 2.42, respectively, and it is unclear whether sorption or biodegradation is the dominant removal mechanism. A significant acclimation period was not observed for phenylphenol or propylparaben (Figure 5-2), but the acclimation to the compounds could have been immediate or extremely short, on the order of minutes. Butylated hydroxyanisole was partially removed by 30% and 40% for SRT values of 5 and 10 days, respectively. Butylated hydroxyanisole has a log D value of 3.5 at pH 8.5, but it is unclear whether sorption or biodegradation is the dominant removal mechanism.

Table 5-4. Tendency of Removal by Activated-Sludge Treatment for Studied Organic Compounds under Differing SRT Conditions.

SRT (d)	1	5	10	35	120
Days of operation (d)	38	28-60[1]	42	24	21
Butylated hydroxyanisole[2]	na	+	++	na	na
Carbamazepine	-	-	-	-	-
DEET[2]	-	+++	+++	na	na
Diclofenac	-	-	-	-	-
Dilantin	na	-	-	na	na
Fenofibrate	+++	+++	Na	na	na
Gemfibrozil	-	-	-	na	na
Ibuprofen	+++	+++	+++	+++	+++
Indolebutyric acid[2]	na	+++	+++	na	na
Naproxen	+++	+++	+++	+++	+++
Phenacetine	+++	+++	+++	+++	+++
Phenylphenol[2]	na	+++	+++	na	na
Primidone	-	-	-	-	-
Propylparaben[2]	na	+++	+++	na	na
Propyphenazone	-	-	Na	na	na
Sulfamethoxazole	-	+++	+++	na	na
Triclosan[2]	+++	+++	+++	+++	+++
Trimethoprim	na	-	+	na	na

(-) no removal, (+) low removal, (++) moderate removal, (+++) high removal, (na) - not analyzed

1 - See Figure 5.4 for the number days of operation for an individual compound

2 - Tier 1 or 2 HPV compound

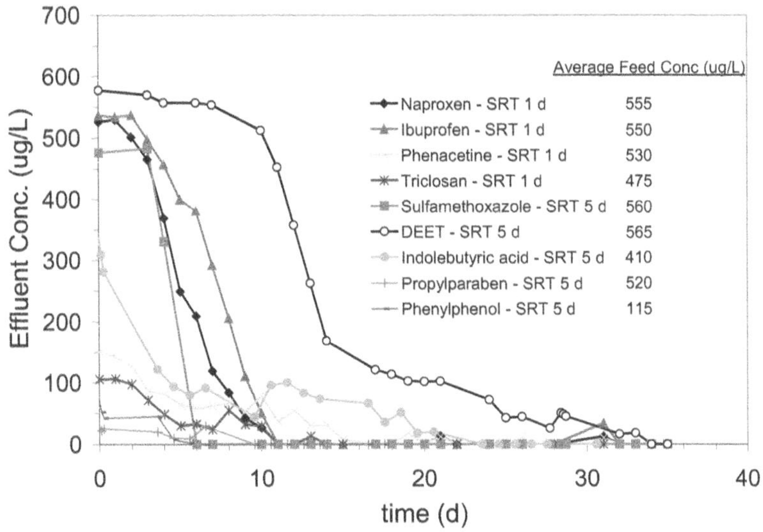

Figure 5-2. Acclimation Periods and Kinetic Removal for Biodegradable Compounds.

The feed solution with spiked compounds was maintained at ambient temperature and was replaced within seven days. Results presented in Figures 5-2 and 5-4 reveal that the compounds remained constant in the feed (new feed vs. old feed) over the course of 7 days with the exception of triclosan and fenofibrate where concentrations decreased in the feed after 7 days. The loss is likely due to sorption to the plastic 50-gallon barrel and/or inlet tubing since both of these compounds had the highest log D values, 4.11, and 5.19 at pH 8.5.

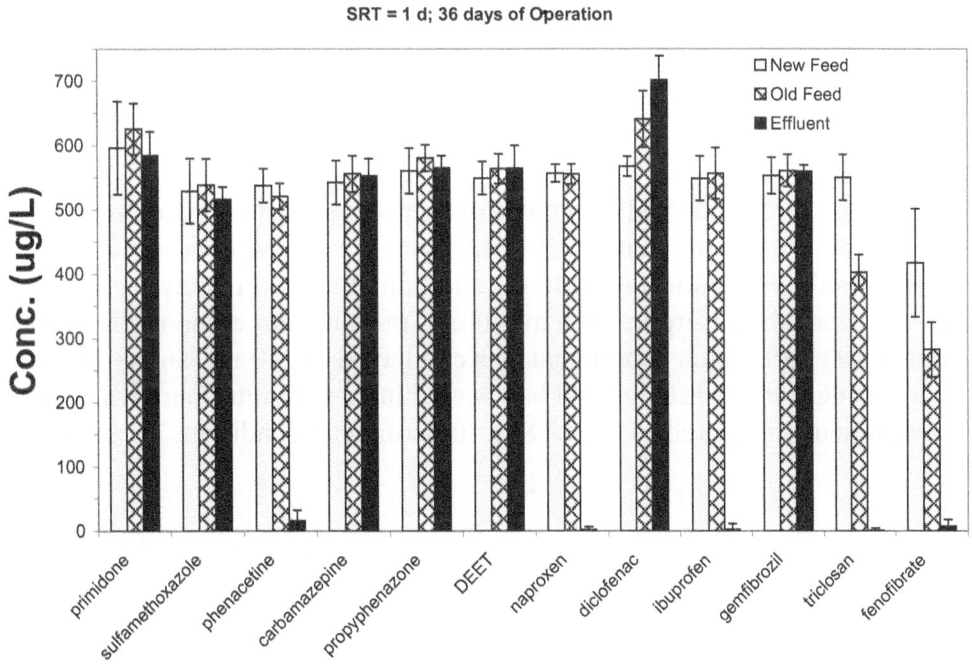

Figure 5-3. Influent and Effluent Concentrations for a SRT of 1 Day. (Note: "New feed" represents the day when a fresh feed was made, where "old feed" represents the feed sample after 7 days).

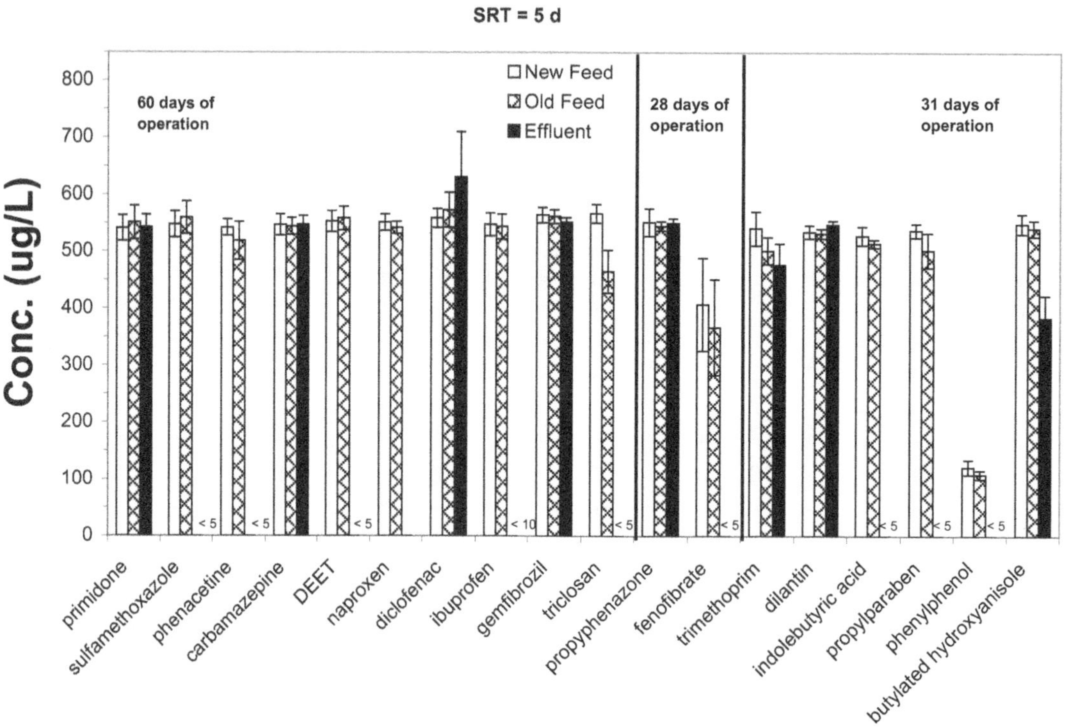

Figure 5-4. Influent and Effluent Concentrations for a SRT of 5 Days. (Note: "New feed" represents the day when a fresh feed was made, where "old feed" represents the feed sample after 7 days).

5.2.1.2 Activated-Sludge Sorption

Bench-scale batch experiments were completed to develop adsorption isotherms for five indicator compounds in an aqueous/inactivated sludge system. The five indicator compounds included phenacetine, carbamazepine, naproxen, diclofenac, and ibuprofen, and their properties are shown in Table 2-4. The indicator compounds have a range of physical and chemical properties similar to many of the Tier 1 and Tier 2 target study compounds, which are also shown in Table 2-4.

Experiments were conducted using ultra-pure water mixed with varying concentrations of activated-sludge collected from Plant 3's fully nitrifying treatment process. The activated sludge was inactivated shortly after collection with the addition of approximately 1 g/L of sodium azide. Experiments examined the adsorption of a mixture of the indicator compounds, each at a concentration of 100 µg/L. Each experiment was conducted in 250-mL amber glass bottles with zero headspace using activated sludge that had been diluted with ultra-pure water to achieve the desired concentrations (measured as mg/LTSS). Experimental conditions are summarized in Table 5-5.

Table 5-5. Inactivated Sludge Experimental Conditions.

	100 μg/L Indicator Compound Concentration									
Inactivated Sludge Concentration (mg/L TSS)	0	300	600	900	1,200	1,800	2,100	2,400	2,700	3,000

Once the 250-mL bottles were filled with ultra-pure water and the appropriate amount of activated-sludge, the experiment was initiated (t_0) by spiking each bottle with a stock solution containing 10 mg/L of each indicator compound for a final concentration of 100 μg/L. The bottles were placed on a shaker table for approximately 24 hours. Water samples were then drawn from each bottle after allowing the sludge to settle to the bottom of the bottle. The samples were filtered using a 0.45-μm polyethersulfone filter (Pall Corporation) and analyzed using HPLC methodology (Section 3.6). A sample was also collected from a separate bottle at t_0 to verify initial concentrations of each compound.

Figure 5-5 shows the removal of each compound as a function of inactivated-sludge concentration. Phenacetine was the only compound demonstrating any significant removal; diclofenac was only slightly removed, while the remaining compounds generally did not appear to be removed. These findings suggest that sorption of these target indicator compounds and similar Tier 1 and 2 compounds onto biosolids is rather small.

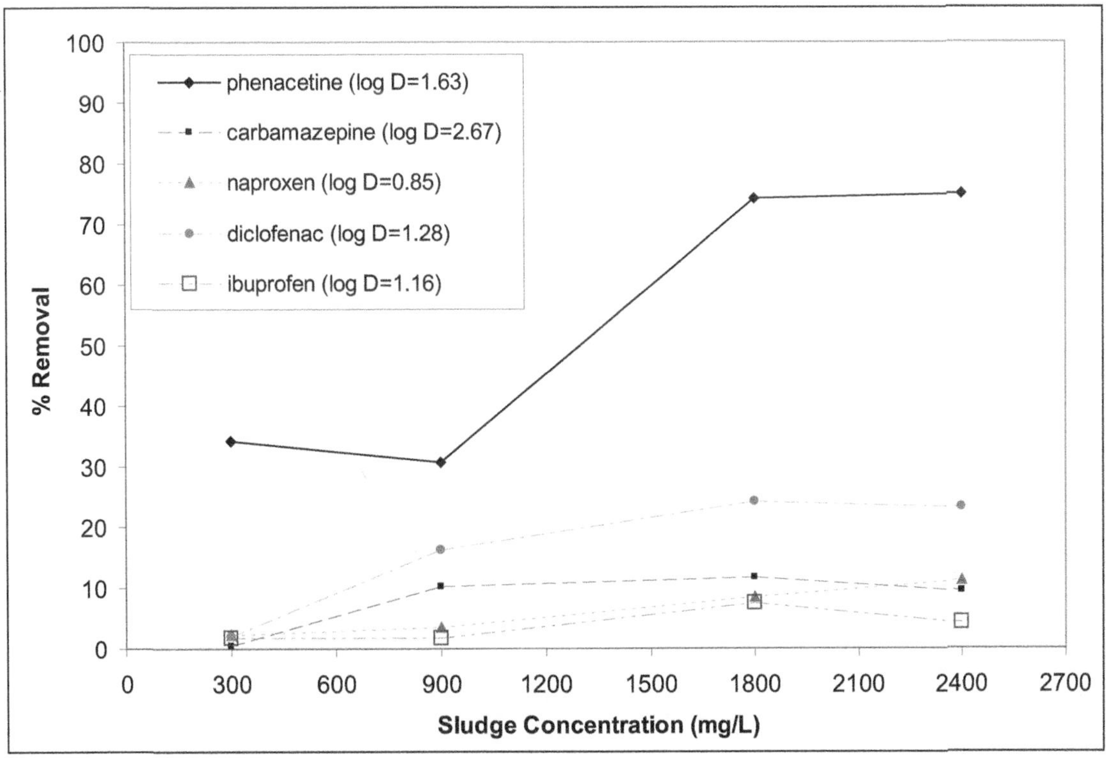

Figure 5-5. Percent Removal as a Function of Sludge Concentration.

5.2.2 Conventional Disinfection Processes

To investigate the efficiency of commonly employed wastewater disinfection processes, samples were collected from the tertiary treated effluent and final (post-disinfection) treated effluents at three full-scale plants (Plants 5, 6, and 7) in order to evaluate the potential treatment effectiveness of different disinfection processes. Plants 5, 6, and 7 use UV disinfection (low pressure, low output lamps), chloramination, and chlorination (sodium hypochlorite), respectively. Concentrations before and after disinfection for Plants 5, 6, and 7 are summarized in Table 5-6. Concentrations of the target compounds detected in the tertiary effluents (before disinfection) were generally variable between the three processes. Removal efficiencies for a select number of Tier 1 and 2 compounds are compared in Figure 5-6. For most compounds chlorination was the best for removing compounds, followed by chloramination, and UV disinfection achieved little or no removal of compounds. Chlorination can transform organic compounds, particularly compounds with electron-donating groups and/or carbon-carbon double bonds via oxidation, substitution, and addition reactions.

Table 5-6. Comparison of Different Full-Scale Disinfection Processes (Plants 5, 6, and 7).

Analyte[1]	Plant 5 Tertiary Effluent	Plant 5 Final Effluent	Plant 6 Tertiary Effluent	Plant 6 Final Effluent	Plant 7 Tertiary Effluent	Plant 7 Final Effluent
Disinfection Type	**UV – Low Pressure**		**Chloramination**		**Sodium Hypochlorite**	
			Tier 1 Compounds (ng/L)			
Atrazine	<5	<5	<1	<1	<1	<1
Hexabromocyclododecane	<100	<100	<100	<100	<100	<100
Dibutyl phthalate	251	179	2,260	1,330	414	212
Bisphenol-A	2,290	2,550	1,210	860	2,260	1,890
Benzophenone	48	71	63	72	45	118
Oxybenzone	<5	<5	2.6	1.1	1.5	<1
BHT	30	35	145	242	49	<25
2-Phenoxyethanol	<100	<100	142	<100	181	<100
Vanillin	464	465	853	293	221	145
Triclosan	36	43	93	18	72	3.4
o-Phenylphenol	<25	<25	22	<10	11	<10
			Tier 2 Compounds (ng/L)			
Acriflavine	<5	<5	<1	<1	<1	<1
Trifluralin	<25	<25	<25	<25	<25	<25
Simazine	<5	<5	<1	<1	<1	1.0
3-Indolebutyric acid	280	285	271	257	233	101
DEET	62	51	205	202	88	82
MGK-11	<25	<25	<25	<25	<25	<25
Hydrocortisone	<5	<5	<1	<1	<1	<1
Camphor	<50	<50	<50	<50	<50	<50
Menthol	<50	<50	<50	<50	<50	<50
2-Methylresorcinol	<100	<100	<100	<100	<100	<100
Propylparaben	<1.25	2.6	0.78	0.26	<0.25	<0.25
Isobutylparaben	<1.25	<1.25	0.50	0.31	0.32	<0.25
BHA	24	26	50	3.3	25	<1
			Model Compounds (ng/L)			
Triclocarban	115	111	127	127	84	26
LAS[3]	na	na	na	na	na	na

Notes: 1 – All values reported were analyzed by LC/MS-MS unless otherwise noted
"<" – value reflects the preliminary reporting limit for that compound
na – not analyzed

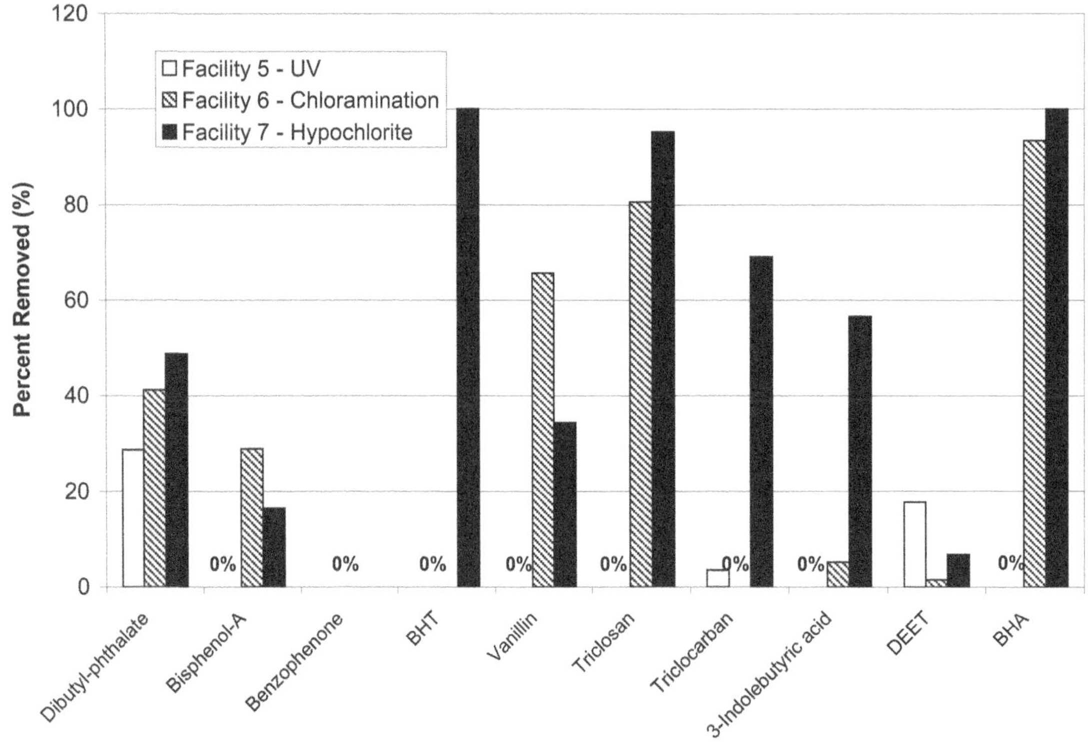

Figure 5-6. Comparison of Different Disinfection Processes for Selected Tier 1 and 2 HPV Chemicals.

5.2.3 Ozonation and Advanced Oxidation Processes

Laboratory-scale oxidation studies were conducted at SNWA's lab-scale ozonation facility. The lab-scale plant design consists of a 12-cell, continuous-flow ozone contactor constructed of inert materials such as glass, fluorocarbon polymers, and stainless steel (Figure 5-7).

Figure 5-7. Ozonation Laboratory-Scale Experimental Set-Up.

The 12 contactor cells each have an HRT of 2 minutes, for a total of 24 minutes. A 55-gallon stainless steel drum was used to feed the tertiary-treated wastewater effluent from Plant 5 at a design flow rate of 1 L/min into the system. The experimental conditions for the experiments are shown in Table 5-7. Ozone was injected into cell 1 with counter-current flow through a fritted glass diffuser producing a bubble size of 10-20 μm. Samples were collected after contact times of 6 (2.1 mg/L of ozone) and 18 minutes (3.6 and 7 mg/L of ozone) where the ozone residual concentration was measured as zero at these contact times. Additional experiments were performed with hydrogen peroxide (H_2O_2) applied in addition to ozone. Experimental conditions with the use of H_2O_2 are also summarized in Table 5-7. Grab samples were collected prior to and during the experiment.

Table 5-7. Ozonation Laboratory-Scale Plant Experimental Conditions.

Experiment #	Ozone Concentration (mg/L)	Hydrogen Peroxide Concentration (mg/L)	Contact Time (min)
1	2.1	-	6
2	3.6	-	18
3	7.0	-	18
4	2.1	1.0	10
5	3.6	2.0	10
6	7.1	3.5	10

Concentrations of the Tier 1 and 2 compounds prior to and after ozonation and ozonation/ H_2O_2 treatments are summarized in Table 5-8. Compounds were not spiked into the tertiary-treated wastewater and the reported concentrations are environmental occurrence concentrations as present in the effluent of Plant 5. The removal percentages of compounds due to ozonation are shown in Figure 5-8. A greater removal percentage was observed as the ozone dose concentration increased (e.g., bisphenol-A, dibutyl-phthalate, 3-indolebutyric acid, and DEET). Similar results were observed with increasing ozone/hydrogen peroxide dose concentrations. In general, when comparing the performance of using ozone alone versus ozone/ H_2O_2 with the same ozone dose concentration, the addition of H_2O_2 did not increase the removal of the compounds (Figure 5-9). However, for 3-indolebuytric acid, the addition of H_2O_2 increased the removal percentage for all three doses. Even though ozonation and ozonation/H_2O_2 treatments use different oxidation mechanisms, they remove the Tier 1 and 2 compounds similarly.

Table 5-8. Compound Concentrations Before and After Addition of Ozone and Hydrogen Peroxide to Tertiary-Treated Wastewater.

Analyte	Analytical Method	Average Tertiary Effluent[1]	2.1 mg/L O3 - 6 min	3.6 mg/L O3 - 18 min	7.0 mg/L O3 - 18 min	Average Tertiary Effluent[1]	2.1 mg/L O3 1.0 mg/L H2O2 - 10 min	3.6 mg/L O3 2.0 mg/L H2O2 - 10 min	7.1 mg/L O3 3.5 mg/L H2O2 - 10 min
Date		1/25/2006	1/25/2006	1/25/2006	1/25/2006	1/26/2006	1/26/2006	1/26/2006	1/26/2006
Tier 1 Compounds (ng/L)									
Atrazine	LC/MS-MS	<10	<1.0	<1.0	<1.0	<10	<1.0	<1.0	<1.0
Hexabromododecane	GC/MS-MS	<25.0	<100.0	<100.0	<100.0	<25.0	<100.0	<100.0	<100.0
Bisphenol-A	LC/MS-MS	2,370	2,520	1,400	508	1,014	1,040	644	113
Dibutyl phthalate	GC/MS-MS	291	218	196	134	188	201	125	87
Benzophenone	GC/MS-MS	47.5	55	<25.0	<25.0	156.5	119	72	<25.0
Oxybenzone	LC/MS-MS	<10	<1.0	<1.0	<1.0	<10	<1.0	<1.0	<1.0
BHT	GC/MS-MS	n.r.	<25.0	<25.0	<25.0	63	<25.0	<25.0	<25.0
2-Phenoxyethanol	GC/MS-MS	<100.0	<100.0	<100.0	<100.0	110.5	<100.0	<100.0	<100.0
Vanillin	GC/MS-MS	223	292	183	165	225	307	200	119
Triclosan	LC/MS-MS	41	1.8	1.6	<1.0	90.5	1.6	1.6	1.9
o-Phenylphenol	LC/MS-MS	<100	<10	<10	<10	<100	<10	<10	<10
	GC/MS-MS	<25	<25.0	<25.0	<25.0	61.5	<25.0	<25.0	<25.0
Tier 2 Compounds (ng/L)									
Acriflavine	LC/MS-MS	<10	<1.0	<1.0	<1.0	<10	<1.0	<1.0	<1.0
Trifluralin	GC/MS-MS	<25.0	<25.0	<25.0	<25.0	<25.0	<25.0	<25.0	<25.0
Simazine	LC/MS-MS	<10	1.2	<1.0	<1.0	<10	<1.0	<1.0	<1.0
3-indolebutyric acid	LC/MS-MS	122.5	108	66	23	197	82	79	19
DEET	LC/MS-MS	92	65	27	3.2	89	60	30	0.96
	GC/MS-MS	111	94	55	32	124.5	108	66	<25.0
MGK-11	GC/MS-MS	<25.0	<25.0	<25.0	<25.0	<25.0	<25.0	<25.0	<25.0
Hydrocortisone	LC/MS-MS	<10	<1.0	<1.0	<1.0	<10	<1.0	<1.0	<1.0
Camphor	GC/MS-MS	<50.0	<50.0	<50.0	<50.0	<50.0	<50.0	<50.0	<50.0
Menthol	GC/MS-MS	<50.0	<50.0	<50.0	<50.0	<50.0	<50.0	<50.0	<50.0
2-Methylresorcinol	GC/MS-MS	<100.0	<100.0	<100.0	<100.0	<100.0	<100.0	<100.0	<100.0
Propylparaben	LC/MS-MS	<2.5	<0.25	<0.25	<0.25	<2.5	<0.25	<0.25	<0.25
Isobutylparaben	LC/MS-MS	<2.5	<0.25	<2.5	<0.25	<2.5	<0.25	<0.25	<0.25
BHA	LC/MS-MS	19	<1.0	<1.0	<1.0	66.5	<1.0	<1.0	<1.0
	GC/MS-MS	<25.0	<25.0	<25.0	<25.0	59	<25.0	<25.0	<25.0
Model Compounds (ng/L)									
Triclocarban	LC/MS-MS	75.5	41	3.3	0.80	112.5	53	19	1.8
LAS[3]	LC-MS	n.r.	n.r.	n.r.	n.r.	n.r.	n.r.	n.r.	n.r.

Notes: 1 –tertiary effluent values reflect an average of 2 values (n = 2); "<" value reflects the preliminary reporting limit; n.r. – not reported

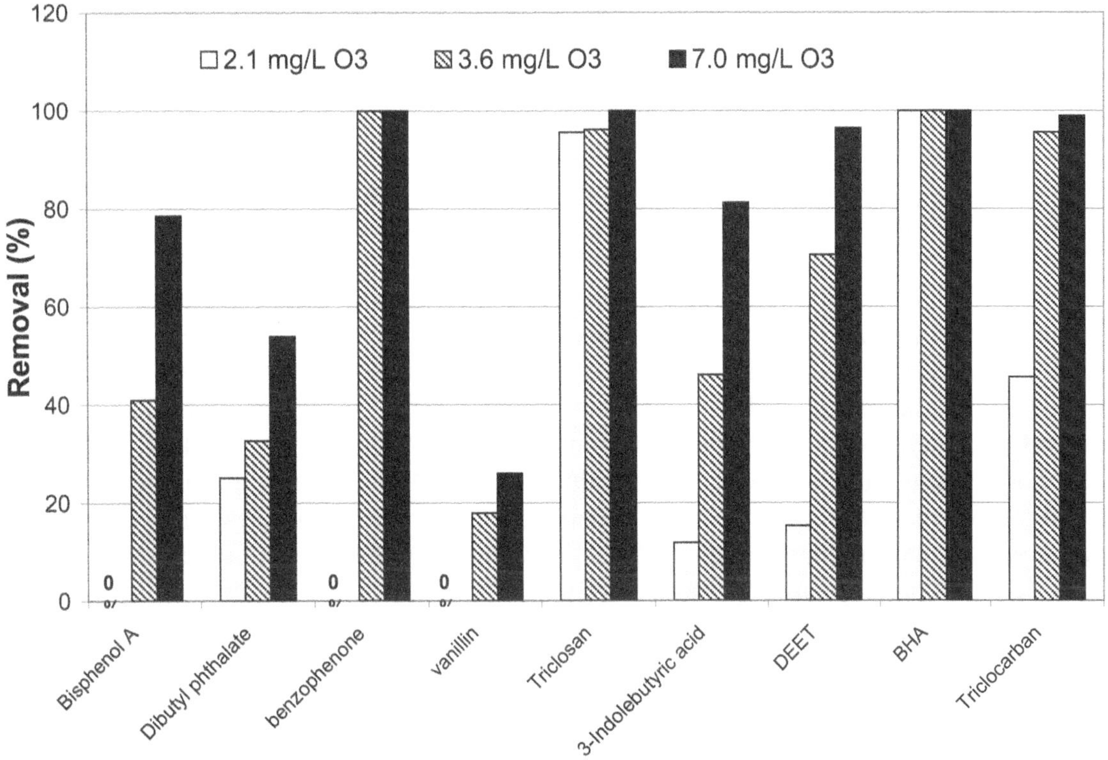

Figure 5-8. Chemical Compound Removal Percentages during Ozonation.

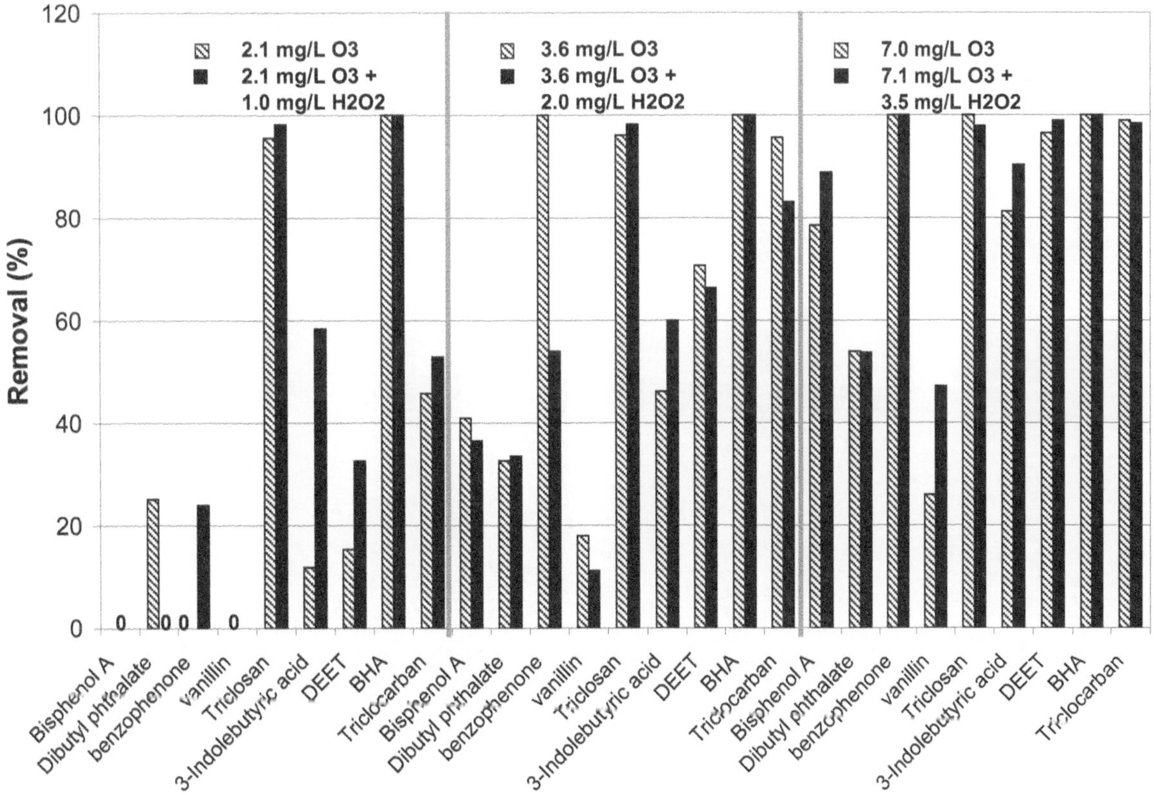

Figure 5-9. Comparison of Compound Removal during Ozonation and Ozonation Coupled with Hydrogen Peroxide.

5.3 Conclusions

The removal of HPV chemicals by biological treatment (activated-sludge and MBR), disinfection (UV, chloramines, chlorine), and advanced oxidation (ozone and ozone/ H_2O_2) processes was examined. In general MBR and activated-sludge treatments removed the Tier 1 and 2 compounds similarly. All detected Tier 1 and 2 compounds were partially or completely removed during biological treatment indicating that these compounds are not both recalcitrant and highly soluble. Biodegradation and sorption are the dominant removal mechanisms during biodegradation treatment. In flow-through laboratory experiments five out of six Tier 1 and 2 compounds were well removed for SRT > 5 days. Butylated hydroxyanisole was only partially removed (40%) at an SRT of 5 days. Triclosan was shown the be effectively removed even at a lower SRT of 1 day; however, another compound, DEET, was not removed at SRT of 1 day, therefore, increasing SRT was beneficial for this compound (the other four compounds were not analyzed at SRT of 1 day). In these same experiments, the removal of target indicator compounds, pharmaceutical residues, was also evaluated. Of these compounds, some were well removed (e.g., ibuprofen, naproxen) and some compounds were not effectively removed (e.g., carbamazepine, primidone, diclofenac) even for SRTs up to 120 days. These later compounds are not representative of the Tier 1 and Tier 2 HPV compounds. The flow-through laboratory experiments also demonstrate that acclimation to some compounds is necessary to achieve effective removal. As Chapter 4.0 discussed, many of the Tier 1 and 2 compounds are observed in the raw water and, thus, there is a constant input of the organic compound to ensure acclimation and biological removal. For conventional disinfection processes, chlorination represents the best process for partial removal of Tier 1 and 2 compounds and a second barrier. Chloramine and UV disinfection processes are not viable removal barriers. Ozonation and ozone/ H_2O_2treatments performed similarly and are additional barrier options where higher effluent quality is desired. However, they do not represent a 100% barrier for all compounds (e.g., vanillin).

CHAPTER 6.0

USING QSPR TO PREDICT THE REMOVAL OF INDIVIDUAL COMPOUNDS THROUGH WASTEWATER TREATMENT PROCESSES

6.1 Introduction

New synthetic organic chemicals are continuously created and it is costly and timely to experimentally study the fate, transport, and toxicology of current and future emerging contaminants on an individual basis. Therefore, it is essential to develop tools that can be used to quickly screen how effectively organic compounds can be removed during conventional and advanced wastewater treatment. Such a tool will provide utilities the ability to quickly screen an organic compound of concern and provide a meaningful response in regard to the fate of the compound and, if needed, provide the degree of treatment upgrade required for removal or recommend source control. QSARs, which have been widely used in developing pharmaceuticals for predicting therapeutic response and assessing the toxicity of organic compounds, have merit for quickly screening the environmental fate of emerging contaminants and *a priori* assessing their removal by treatment systems.

The concept behind QSAR and QSPR analyses is to mathematically quantify the correlations between a fate parameter and molecular descriptors of a compound. These QSAR/QSPR techniques allow for removal predictions for the following major fate and treatment routes within conventional and advanced wastewater treatment: phase partitioning (e.g., air-water Henry's Law constant; aqueous-phase primary solids adsorption; aqueous-phase biosolids adsorption; aqueous-phase activated carbon adsorption); reactivity (e.g., hydrolysis, biodegradation, oxidation), and physical separation (e.g., membrane transport).

The critical step for performing QSAR/QSPR modeling is the choice of significant chemically or structurally meaningful molecular descriptive parameters. This chapter highlights gaps in existing QSAR/QSPR efforts for representative removal mechanisms and specific treatment processes.

6.2 Overview of QSPR Models

The term QSPR has often been used in place of the term QSAR when a fate *property* is desired as opposed to a biological *activity* response. Therefore, the models presented in this chapter will be referred to as QSPR models. The concept of QSPR analysis is to mathematically (usually empirically) quantify the correlations between a fate parameter (e.g., degradation rate constant, partitioning coefficient) and molecular descriptors of a compound (e.g., volume, connectivity indices, dipole moment). Once a correlation is found, the treatability of emerging and current organic contaminants can be predicted using mass balance models.

6.2.1 Development of QSPR Models

The principal elements involved in the development of a QSPR model are illustrated in concept in Figure 6-1. The first step in the model development is selecting a training set of compounds. In general, a QSPR model will be more robust with a larger training dataset. The second step is the attainment of environmental fate parameters, preferably experimentally, such as sorption partitioning and removal rate constants. The third step is the attainment of suitable structural descriptors influencing the removal mechanisms. Common molecular descriptors are listed in Table 6-1, which can be determined theoretically, empirically, or experimentally.

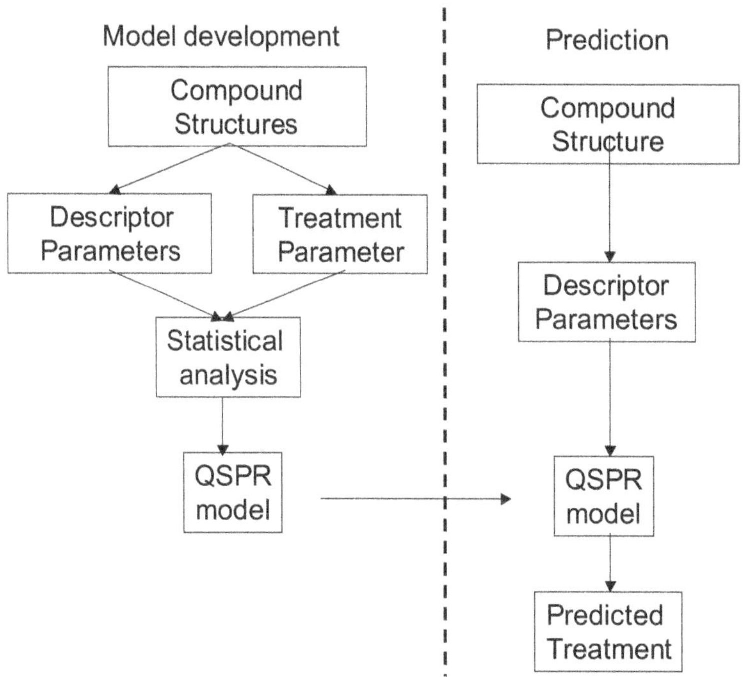

Figure 6-1. Principal Elements Involved in the Development of a QSPR Model.

Table 6-1. Common Organic Compound Molecular Descriptors.

Classes	Parameters
Constitutional	molecular weight, number of rings, number of fragments, number of functional groups
Topological	connectivity indices
Electrostatic	polarizability, dipole moment
Geometrical	molecular volume, surface area, van der Waals radius, Taft constants
Statistical Mechanical	vibrational frequencies, thermochemical parameters
Quantum Chemical	orbital energies, atomic charges

Classical descriptors are often determined using the 2D representation of the molecule and other descriptors require the generation of 3D structures. Nowadays, 3D-based molecular descriptors, such as quantum-mechanical and -chemical descriptors, can be calculated relatively quickly due to the recent advancement of computational hardware and semi-empirical methods. Thus, quantum-mechanical and -chemical calculations are an attractive source for novel molecular descriptors that take into account all the electronic and geometrical interactions within a molecule. It is important to note the variability inherent in the measured or calculated fate parameters and descriptors will determine the variability in the final prediction. The fourth step is the application of statistical methods (e.g., multilinear regression models, partial least squares, principal component regression, artificial neural networks) to derive correlations between the fate parameter and structural descriptor data sets. The fifth step is the validation of the QSPR model using another set of data outside the training data. Finally, the application boundary conditions for the model need to be defined (e.g. environmental matrix, operational conditions). The model development process is usually iterative, that is, a number of QSPR models are generated using different methods and parameters. During this step, it is extremely important to choose relationship(s) with molecular descriptors that are chemically and/or structurally meaningful to the respective removal mechanism. The results are compared and the most predictive and meaningful model(s) are selected for future deployment. Before the models can be applied for prediction, the appropriate molecular descriptors need to be determined for the organic compound of interest. For a successful and *applicable* QSPR model, the descriptors need to be easily measurable or calculable.

6.3 Existing QSPR Techniques

The following section provides a summary on relevant QSPR molecular descriptive parameters used in predictive models for the following major fate and treatment routes of various organic compounds: reactivity, phase partitioning, and physical separation.

6.3.1 Reactivity

Within water reclamation unit processes, organic compounds may be reactively transformed or degraded by chemical oxidation, hydrolysis, or biodegradation mechanisms.

6.3.1.1 Oxidation

Organic compounds can undergo chemical oxidation, such as in ozonation, chlorination, and advanced oxidation processes. Organic compounds with electron donating groups and/or carbon-carbon double bonds are favored for oxidation to occur. The Hammett constant, molecular orbital energies (high and low), polarity (dipole moment), and oxidation potential are the most commonly used molecular descriptors for QSPR analysis for the reactivity of organic compounds during oxidation processes under laboratory conditions. Most studies developed QSPRs that predicted degradation rate constants via ozonation (Gould, 1987; Hoigne and Bader, 1983; Hu et al., 2000), chlorine dioxide (Tratnyek and Hoigne, 1994), advanced oxidation processes (e.g., Fenton's reaction, UV/H_2O_2, UV/O_3, UV/TiO_2) (Tang 2004; Cooper and Cunningham, 1992; Percival et al., 1995; Bartolotti and Edney, 1994). Previous studies have demonstrated that chlorine disinfection represents an effective barrier for many trace organics present in wastewater effluents (Westerhoff et al. 2005). However, limited QSPRs exist for chlorine disinfection; therefore there is a need to develop QSPRs to predict chlorination degradation rates. Note, chlorination reactions are a combination of three reactions: oxidation, substitution, and addition. A recent study developed 3D-based QSPR models for predicting the percent removal of a compound by ozone and chlorine (Lei and Snyder, 2007). These models found the π surface area, number of bonds amenable to electrophilic attack, and ionization potential to be important molecular descriptors. However, because these models predict "percentage removal," they are bound by the experimental conditions of the training set (e.g., ozone dose, contact time). Most oxidation QSPR models have been developed from simple laboratory-controlled experimental matrices; therefore, the QSPR models are perhaps less effective for realistic complex water matrices. The presence of dissolved inorganic compounds and organic matter may result in differing kinetic rate constants due to competitive and synergistic effects. Therefore, the bench-scale QSPR models need to be further validated at field-scale conditions.

2D-based QSPRs are available for the hydroxyl radical reactions with organic compounds in the gas phase (Atkinson, 1986, 1987, 1988; Kwok and Atkinson, 1995). It is plausible that hydroxyl gas-phase QSPRs can be applied to the aqueous environment. Haag and Yao (1992) correlated hydroxyl radical H-atom abstraction rate constants for substituted alkanes in water vs. the gas phase. Similar chlorination QSPRs could be generated as data are available on the chlorination of organics in air (Blanco et al., 2007).

6.3.1.2 Hydrolysis

Hydrolysis is the reaction of organic compounds with water, and it is a function of the pH of the system. Thus, a compound can potentially hydrolyze if the pH changes within a treatment system. Peijnenburg (1991) presented a large number of QSPR models for the prediction of hydrolysis rate constants under laboratory-scale conditions. The hydrolysis rate constants were correlated with molecular descriptors that consisted of the Hammett constant, a steric contribution parameter, and pK_a.

6.3.1.3 Biodegradation

Many organic compounds can be degraded by microorganisms, such as those existing in activated sludge or soil aquifer treatment systems. A couple of QSPR models, sometimes referred to as quantitative structure biodegradability relationship (QSBR) models, have been proposed for predicting organic compound biodegradation (e.g., biodegradation rate constant, BOD) using various molecular descriptors (Banerjee et al., 1984; Boethling 1986; Dearden and Nicholson, 1986, 1987; Parsons et al., 1991; Peijnenburg 1994; Dearden and Cronin, 1996; Okey and Stensel, 1996; Raymond et al., 2001). The molecular descriptors commonly used in these models represented connectivity indexes, chemical groups, electronic (e.g., superdelocalizability) and steric (e.g., van der Waals radius) parameters.

Several *qualitative* structural property relationship (SPR) models have been developed that predict the probability of the biodegradability of organic chemicals (Niemi et al., 1987; Howard et al., 1992; Klopman et al., 1993; Dearden and Cornin, 1996; Loonen et al., 1999; Hiromatsu et al., 2000; Sabljic and Peijnenburg, 2001; Raymond et al., 2001; Boethling et al., 1994; Roriji et al., 2002; Alikhanidi and Takahashi, 2004). For these studies, the functional group and/or substructure (molecular fragments) of the chemical compounds are used to determine the tendency of biodegradation of an organic compound.

Direct methods for predicting rates of biodegradation are lacking because of the complexity of the biodegradation process. Biodegradation rates are not only affected by the compound's structure, but also by acclimation and the environmental conditions. Most biodegradation QSPR models were developed from simple laboratory-controlled experimental matrices, and some studies utilized conditions and biomass (e.g., concentration of test compound and inoculum) that may not be representative of field-scale scenarios (e.g., soil-aquifer treatment vs. activated sludge). Biodegradation testing should be performed under environmental conditions in which the chemical is expected to reside. Therefore, reliable QSBR methods are limited due to the lack of uniformly measured biodegradation data.

6.3.2 Phase Partitioning

Within water reclamation processes organic compounds may be removed by phase partitioning via the air-aqueous, soil-aqueous, biosolids-aqueous, or activated carbon-aqueous interfaces.

6.3.2.1 Volatilization

Volatile soluble organic compounds can partition into the air phase, such as in aeration basins. The Henry's law constant is the equilibrium distribution of a chemical between the air and water phases. Robust QSPR predictive models have been developed for predicting Henry's law constants (Sabljic and Güsten, 1989; Nirmalakhandan and Speece, 1988). The models utilized fragment group, polarizability, connectivity indexes, and/or hydrogen bonding molecular descriptor parameters.

6.3.2.2 Soil Adsorption

Hydrophobic organic compounds can sorb onto soil surfaces, such as those present in soil aquifer treatment systems. More than 200 QSPR models have been developed, some of which are robust, for the estimation of soil partitioning coefficient for diverse organic compounds. These models are presented by Gawlik et al. (1997) and Kahn et al. (2005). The soil-partitioning coefficient was found to be best predicted using the first-order connectivity molecular descriptor (Sablijic, 1987; Sabljic et al., 1995). Most soil adsorption QSPR models have been developed from simple laboratory-controlled experimental matrices, and the conditions and soil type used may not be represented of field-scale scenarios. Therefore, the bench-scale QSPR models need to be further validated for field-scale conditions.

6.3.2.3 Biosolids Adsorption

Hydrophobic organic compounds can also potentially sorb onto biosolids, such as those in activated sludge processes. The physical, chemical, and structural properties of biosolids differ from soils, which can affect the organic chemical adsorption behavior. Only a few studies of organic compound sorption on to wastewater solids have been reported (Dobbs et al., 1989; Greaves et al., 2001; Hsieh and Mukherjee, 2003). The models utilized fragment group and connectivity index molecular descriptor parameters. Most of the biosolids adsorption QSPR models were developed from simple laboratory-controlled experimental matrices and the conditions may not be represented of field-scale scenarios. Therefore, the bench-scale QSPR models need to be further validated for field-scale conditions.

6.3.2.4 Activated Carbon Adsorption

Hydrophobic organic compounds can sorb onto activated-carbon surfaces, such as those in powdered or granular activated carbon processes. Some studies have developed QSPR models for predicting organic compound removal via activated carbon adsorption using molecular descriptors (Blum et al., 1994; Brasquet et al., 1997; Brasquet and Cloirec, 1999). The models utilized connectivity index molecular descriptor parameters. The activated carbon adsorption QSPR models were developed from simple laboratory-controlled experimental matrices and the conditions and activated carbon type may not be representative of field-scale scenarios.

Therefore, there is a clear need to validate bench-scale derived QSPR models under field-scale conditions.

6.3.3 Physical Separation

Within water reclamation processes, organic compounds can be removed by membrane rejection. Limited studies have developed QSPR models for predicting the fate of organic chemicals through membranes (Rodriguez et al., 2004; Bellona, 2007). Rodriguez et al. (2004) examined endocrine disrupters, antibiotics, pesticides, and neuroactive drugs and developed QSPR models for predicting the percentage passage through, adsorption to, and rejection via reverse osmosis (RO) membranes of these organic compounds. Because of the various solute-membrane interactions, separate QSPR rejection models were developed for specific membrane types (polyamide and cellulose acetate). The behavior of the organic compounds was found to be correlated to the following molecular descriptor groups: charge/polarity (i.e., dipole moment), molecular complexity (formula weight), hydrophobicity, and hydrogen bonding. A simple experimental matrix was used, and these QSPR models are perhaps less effective for more complex water matrices and different membrane fouling types. An organic compound's rejection behavior can be strongly influenced by the nature of the water matrix and fouling layer. In addition, these models predict percentages, thus making these empirical-based models are highly dependent upon the experimental conditions (e.g., membrane type, operating pressure). Bellona (2007) developed and incorporated a QSPR into a hydrodynamic model for predicting the rejection of nonionic organic compounds by nanofiltration membranes. For this QSPR, an orientation angle-fitting parameter is dependent on the compound's molecular length, and total hydrophobic and hydrophilic portions of the solvent accessible surface areas. The idea is that a solute's rejection is dependent upon a unique orientation angle as a solute approaches the membrane, which, in turn, depends upon the solute structure and properties. However, the orientation angle is also likely a function of the membrane properties, and the QSPR is not valid for other membrane types. These QSPR models derived from bench scale need to be further validated for full-scale conditions.

6.3.4 Gaps in QSPR Models

Several important research gaps were revealed in existing QSPR modeling efforts. Many existing models are restricted only to narrow classes of chemicals and are likely not applicable to organic compounds of widely varied structures (e.g., wastewater-derived chemicals). There is a need to evaluate QSPR models for their applicability to a wide range of organic compounds. Also, most of the models are experimental matrix specific (e.g., soil, biomass and activated-carbon type), and most of the models have been developed from simple laboratory-controlled experimental matrices. Therefore, the models need to be evaluated for more realistic complex water matrices (e.g., dissolved inorganic and organic matter effects) and different biomass, soil, activated carbon types, and membrane fouling types. Also, the accuracy of 2D- and 3D-based QSPRs need to be evaluated and compared, and there is a need to generate fate parameter data (e.g., degradation rate constants, partitioning coefficients) for the evaluation of existing and development of new applicable QSPR models. Finally, QSPR models need to be incorporated into mass balance models and these mass balance models (e.g., fugacity models) need to be validated by full-scale treatment data.

6.4 Computer Software for QSPR Modeling

There are many computer programs available that use QSPR models or can be used to develop QSPR models. The types of programs can be divided into two groups. First, there are programs that calculate compound descriptors using either a 2D (e.g., SMILES notation) or 3D (molecular modeling) representation of the compound, which can be externally correlated with treatment parameters using statistical programs for the generation of QSPR models. Some of these programs incorporate QSPR models that predict fate parameters, such as the Henry's law constant. Secondly, there are programs that can calculate compound descriptors and fate parameters as well as develop QSPR models. These programs have the ability to generate descriptors, accept input treatment parameter data, and then statistically examine the relationships. Essentially, these programs have a statistical package embedded in them. Table E-1 (Appendix E) summarizes the programs that are only used for predicting descriptor and/or fate parameters, and Table E-2 lists programs that can also generate descriptor and/or fate parameters, but can also be used for QSPR development. Public domain programs consist of SPARC (online), mitools (online only), ChemIDPlus, EPI Suite, and QSAR. QSAR is an ongoing public domain project that will enable the generation of compound descriptors and the building of QSPR models.

6.4.1 U.S. EPA's EPI Suite

The EPI (Estimation Programs Interface) Suite v.3.2 Package was chosen to be examined more closely since this program is a public domain software and it contains predictive QSPR models that estimate physical/chemical fate properties of an organic compound, such as its octanol-water partition coefficient, soil adsorption coefficient, hydrolysis rate constant, Henry's Law constant, melting point, boiling point, vapor pressure, water solubility, and biodegradability in activated sludge systems. EPI Suite's fate properties are estimated using atom, group, and/or bond fragment methods based upon Simplified Molecular Input Line Entry System (SMILES) notation of the compound. A SMILES notation depicts a molecular structure as a two-dimensional picture as if drawn on a piece of paper. Fragment methodology consists of a structure that is divided into fragments (atom or larger functional groups), and coefficient values of each fragment or group are summed together to yield the fate property. EPI Suite utilizes these estimated properties to predict the partitioning of chemicals between air, soil, sediment, and water under steady-state conditions for a default model "environment" and the removal of an organic compound during wastewater treatment, where percentage values are given for the total removal and three contributing factors: biodegradation, sorption to sludge, and volatilization. EPI Suite uses a version of the Sewage Treatment Plant (STP) model developed by Clark et al. (1995) and uses the following chemical properties and fate parameters as inputs to the model: molecular weight, water solubility, vapor pressure, log K_{ow}, and biodegradation half-life. Figure 6-2 compares removals of household chemicals observed in a full-scale conventional activated sludge treatment plant (nitrifying/denitrifying) and estimated removals from EPI Suite. For half the compounds, EPI Suite did fairly well in predicting their removals. However, for some compounds, such as camphor, bisphenol A, oxybenzone, hydrocortisone, and DEET, EPI Suite underestimated removal. This discrepancy could be due to the fate properties (e.g., biodegradation half-life) not being well predicted based upon the 2D structure of the compound. The biodegradation rate constant and its dependence on biomass concentration is the most critical and uncertain variable within the STP model. 3D based QSPR techniques could be better indicators for biodegradation potential. Also, EPI Suite is based solely on aerobic biodegradation mechanisms and does not take into account potential removal of compounds under other microbial conditions, such as denitrification.

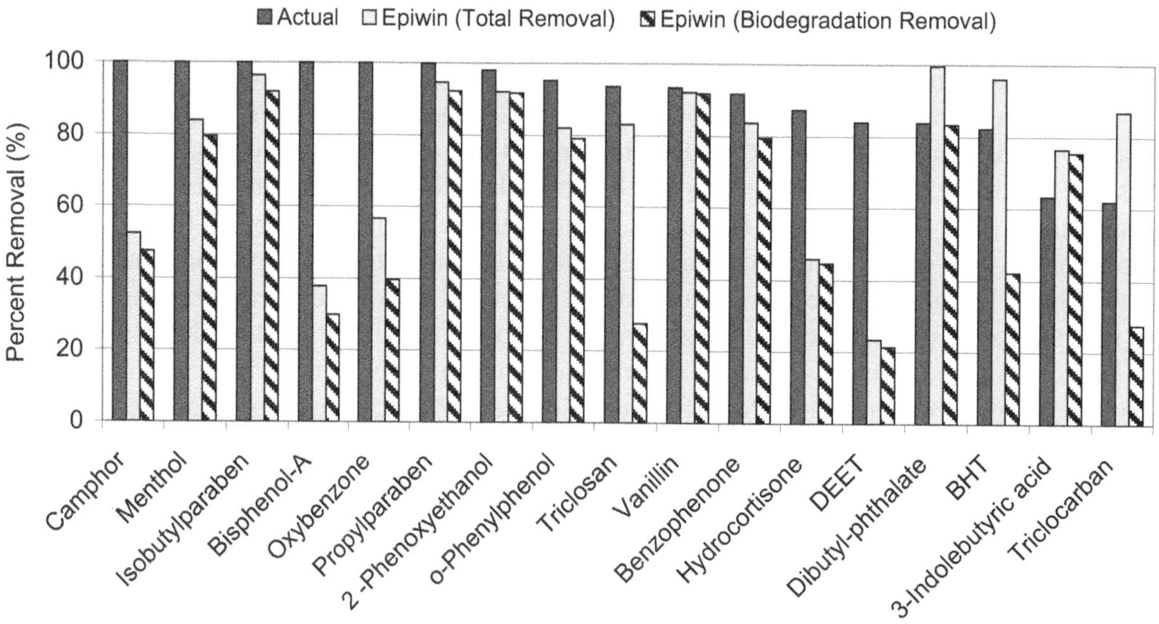

Figure 6-2. Observed Removal of HHCs during Full-Scale Conventional Activated Sludge Treatment and Estimated Removal by EPA Epiwin.

6.4.2 3D-Molecular Modeling Software

In general, 3D-based molecular modeling programs calculate descriptors and properties based on the entire 3D-molecular structure of a compound instead of based on fragment information which is typically used when the 2D structure is only known. 3D-based descriptors, such as dipole moment, total solvent accessible surface area, volume, and solute-as-donor hydrogen bonds, are arguably physically more meaningful than 2D-based descriptors from atom/fragment/bond contribution methods. Schrödinger is an example of a rather expensive 3D-based molecular modeling program. This program is able to generate molecular descriptor parameters based upon the full 3D molecular structure, rather than on fragment information.

6.4.2.1 Comparison of pKa Values

Three models were evaluated for predicting pKa. Schrödinger uses 3D-based quantum mechanical approaches, whereas CHEMID and SPARC (public domain programs) use the 2D representation of the compound to predict the acid dissociation constant (Table 6-2). A list of 11 compounds was selected based on the availability of experimental pKa values and structural features. SPARC seems to be ineffective in predicting the acidity for compounds with amine groups. For example, dilantin is an acid due to the dissociation of the proton from its secondary amine group. The predicted pKa by SPARC is off by a factor of 10. The predictions from Schrödinger and ChemID are in general agreement with the experimental data. This indicates that for this set of compounds the public domain program ChemID is just as good as the rather expensive Schrödinger program for predicting pKa values.

Table 6-2. Comparison of Estimated pKa with Experimental Data.

	Experimental	Schrodinger	ChemID	SPARC
Ethinyl Estradiol	10.4	10.8	9.43	10.38
Triclosan	7.9	8.7	7.98	7.99
Trimethoprim	7.12	7	7.16	6.3
Acetaminophen	9.38	9.6	9.46	9.67
Diclofenac	4.15	4.9	4	4.09
Ibuprofen	4.91	4.9	4.85	4.52
Naproxen	4.15	5	4.2	4.46
Diazepam	3.4	2.7	2.92	2.4
Dilantin	8.33	8.3	9.13	-1.83
Fluoxetine	8.7	9.6	10.3	9.15
Atrazine	1.7	1.6	2	1.11

6.4.2.2 Comparison of Estimated Octanol/Water Partition Coefficients

The octanol/water partition coefficient, log K_{ow}, is often used to assess the hydrophobicity of a compound and its potential partition to solid surfaces. The log K_{ow}, describes the distribution between octanol and water after equilibration for a neutral organic compound. Often the log K_{ow} is predicted by using 2D-based molecular descriptors such the structural fragments of the compound. CLOGP (BioLoom), KOWINN (EPI Suite), and log P DB (ACD) are the most commonly used 2D-based QSPR programs for predicting log K_{ow}. Table 6-3 compares the log K_{ow} values estimated by various programs: ACD, EPI Suite, BioLoom, Dragon, Hyperchem, mitools, and Schrödinger. Results of this exercise demonstrate that the log K_{ow} varies among the various estimation programs. Even EPI Suite was comparable to values generated by Schrödinger, which uses 3D-based molecular descriptors to predict log K_{ow}. The standard deviation ranged from 0.14 to 0.85 log units. EPI Suite was able to predict log K_{ow} values that were similar to the experimental values. Log D was also determined using ACD and EPI Suite. Log D is the partitioning coefficient at pH 7, which is more important than log K_{ow} for charged compounds present at environmental pH near 7. EPI Suite was not able to predict the log D at pH 7 for salicyclic acid and dimethylamine (DMA), for which log D is equal to log K_{ow}. These organic compounds are charged at pH 7, and charged compounds partition differently than neutral species, as shown by the ACD predictions for salicyclic acid and DMA. The estimation programs, with the exception of ACD, did not have the capability to predict the log D for a charged compound at pH 7.

6.4.2.3 2D- vs. 3D-Based Descriptors

Molecular descriptors have been classically derived from the 2D structure (e.g., connectivity indices, group contribution methods), but novel molecular descriptors can be derived from the 3D structure, such as quantum-mechanical and -chemical descriptors (e.g., dipole moment, orbital energies). Log K_{ow} results above demonstrated that 2D- and 3D-molecular modeling programs were comparable in predicting log K_{ow}. Therefore, in some cases 2D-based descriptors are sufficient in predicting a desired treatment parameter, however 2D-based descriptors may not be sufficient to predict a certain treatment parameter. 3D-molecular modeling programs can generate unique molecular descriptors that take into account the electronic and geometrical interactions within a molecule that 2D molecular modeling cannot predict. These 3D-based molecular descriptors provide a much broader scope for QSPR analysis and an enormous potential for predicting the removal or fate of individual compounds through wastewater treatment systems. For example, critical descriptor parameters for oxidation (e.g., molecular orbital energies) and activated carbon adsorption treatment (e.g., dipole moment) are best estimated using 3D molecular modeling programs. Section 6.5 demonstrates that 3D-based descriptors were better in predicting activated-carbon adsorption constants than conventional 2D based descriptors such as log K_{ow} and water solubility.

Table 6-3. Comparison of Estimated Log K_{OW}.

Structure	Chemical				Exp.	ACD	ACD	EPI Suite	EPI Suite	BioLoom Clog P	Dragon Moriguchi	Dragon Ghose-Crippen	Hyperchem	mitools	Schrodinger	Average	Standard deviation
					log K_{OW}	log K_{OW}	log D (pH 7)	log K_{OW}	log D	log K_{OW}	log K_{OW}	log K_{OW}	log K_{OW}	log K_{OW}	log K_{OW}	log K_{OW}	
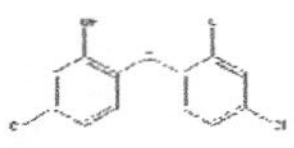	triclosan	hydrophobic	aromatic	uncharged	4.76	5.80	5.75	4.66	4.66	5.53	3.84	5.11	5.03	4.89	4.37	4.89	0.59
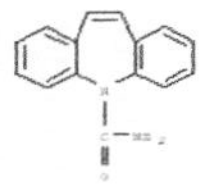	carbamazepine	transphilic	aromatic	uncharged	2.45 2.25 2.19	2.67	2.67	2.25	2.25	2.38	2.59	3.32	2.72	3.96	2.28	2.77	0.59
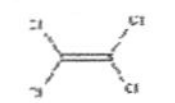	tetra-chloroethylene	transphilic	aliphatic	uncharged	3.40	2.95	2.95	2.97	2.97	3.48	2.50	2.43	2.52	3.35	2.71	2.92	0.41
	farnesol	hydrophobic	aliphatic	uncharged	-	5.31	5.31	5.77	5.77	5.00	3.90	4.76	3.94	5.23	4.52	4.80	0.66
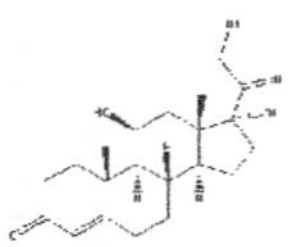	hydrocortisone	hydrophilic	aromatic	uncharged	1.61	1.43	1.43	1.52	1.62	1.89	too many atoms		2.37	1.45	1.56	1.70	0.33
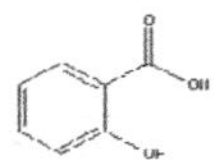	salicylic Acid	hydrophilic	aromatic	charged (-)	1.19	2.26	-1.68	2.24	2.24	2.19	1.64	1.17	3.54	1.36	0.93	1.84	0.82
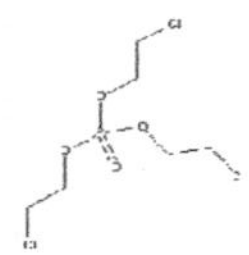	tris-(2-chloroethyl)-phosphate	hydrophilic	aliphatic	uncharged	1.44	0.48	0.48	1.63	1.63	0.47	2.36	2.20	2.70	2.06	2.65	1.78	0.85
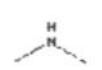	DMA	hydrophilic	aliphatic	charged (+)	-0.38	-0.43	-3.44	-0.17	-0.2	-0.52	-0.17	-0.22	-0.2	-0.12	-0.27	-0.28	0.14

6.5 Derivation of Descriptors through Controlled Laboratory Studies

6.5.1 Powdered Activated Carbon Adsorption

3D-based molecular descriptors were derived for the removal of organic compounds during activated carbon adsorption experiments. Bench-scale batch experiments were conducted to develop adsorption isotherms for indicator compounds in an aqueous/powdered activated carbon (PAC) system. These experiments evaluated the adsorption of six indicator compounds under two primary matrix conditions. Compounds studied are listed in Table 6-5 and their properties are listed in Table 2-4. One matrix consisted of ultra-pure water with varying concentrations of PAC (F300, Calgon Carbon). The second matrix condition consisted of effluent organic matter (EfOM) at a concentration of 5 mg/L with the same type and varying concentrations of PAC. The EfOM used in all experiments was derived from concentrated secondary treated effluent collected from a municipal wastewater treatment facility local to CSM. The experiment evaluated a mixture of the indicator compounds, each at a concentration of 100 µg/L. 250-mL head-space-free amber bottles were used as the reactors. PAC concentrations ranged from 0 to 50 mg/L, which was dosed using a stock slurry of 1,000 mg/L. Experimental conditions are summarized in Table 6-4.

Reactor bottles were filled with the respective matrix solutions (with and without EfOM) and at the initiation of each experiment (t_0), bottles were spiked with a stock solution containing 10 mg/L of each indicator compound for a final concentration of 100 µg/L. The bottles were placed on a shaker table for up to 24 hours to ensure that steady-state conditions were achieved. Analytical samples were collected after 24 hours. The samples were filtered using a 0.45-µm polyethersulfone filter (Pall Corporation), and analytes were quantified using a HPLC method (Section 3-6). The data generated was then used to develop Freundlich adsorption isotherms and determine suitable adsorption coefficients. The researchers used the general Freundlich adsorption equation (Equation 1) to solve for the variables log K_F (adsorption capacity) and 1/n (adsorption strength), where log q_e reflects the compound's concentration adsorbed onto solids and C_e reflects the compound's equilibrium concentration in solution.

$$\log q_e = \log K_F + (1/n)\log C_e \qquad (6.1)$$

Table 6-4. Powdered Activated Carbon Experimental Conditions.

DI only, 100 µg/L Indicator Compound Concentration								
PAC Concentration (mg/L)	0	0.5	1	3	5	10	20	50
5 mg/L Effluent Organic Matter (EfOM), 100 µg/L Indicator Compound Concentration								
PAC Concentration (mg/L)	0	0.5	1	3	5	10	20	50

In the presence of EfOM, compounds were not as readily removed from solution at the higher PAC concentrations. Table 6-5 summarizes the Freundlich adsorption isotherm constants measured in ultra-pure water and ultra-pure water containing effluent organic matter six compounds studied. The Freundlich adsorption isotherm constants were correlated with molecular descriptors estimated from the 3D molecular modeling program, Schrödinger. The following QSPR models were derived:

Ultra-pure water:

Log K = 0.35 (Hydrogen Bond Acceptor) + 0.012 (Volume) – 8.9
($n = 5$, $R^2 = 0.88$; P factor = 0.12; F factor = 7.2)

Effluent organic matter:

Log K = 0.15 (Volume) + 4.34(Dipole Moment) - 3.85(Polarizability) - 28.4
($n = 6$, $R^2 = 0.78$; $P = 0.31$; $F = 2.3$)

The variables/descriptors were found to be independent of one another and significant based upon the P and F factors. Also, only log K needs to be predicted since log K correlates with 1/n (Figure 6-3).

Log D (pH 7) and water solubility were considered in deriving the above models; however, these parameters did not correlate well with the Freundlich activated-carbon adsorption isotherm constants. This bodes for the use of 3D molecular modeling programs to calculate 3D-based descriptors, such as dipole moment for the polar uncharged and charged compounds such as those studied here.

Table 6-5. Freundlich Activated-carbon Adsorption Isotherm Constants in Ultra-pure Water and Ultra-pure Water Containing Effluent Organic Matter (5 mg/L) for Six Studied Compounds.

	Ultra-pure water		**Deionized water Ternes et al. (2002)**		**EfOM (TOC = 5 mg/L)**	
	1/n	Log K log[(mg/g) (L/mg)1/n]	1/n	Log K log[(mg/g) (L/mg)1/n]	1/n	Log K log[(mg/g) (L/mg)1/n]
Primidone	0.46	1.7	-	-	5.8	9.1
Ibuprofen	0.36	1.6	-	-	8.5	12
Diclofenac	0.65	2.1	0.19	2.1	5.3	8.0
Carbamazepine	0.14	1.2	0.38	2.6	1.9	3.3
Naproxen	0.25	1.4	-	-	1.8	2.7
Phenacetine	7.29	13	-	-	4.3	8.7
Clofibric acid	-	-	0.25	1.9	-	-
Bezafibrate	-	-	0.19	2.1	-	-

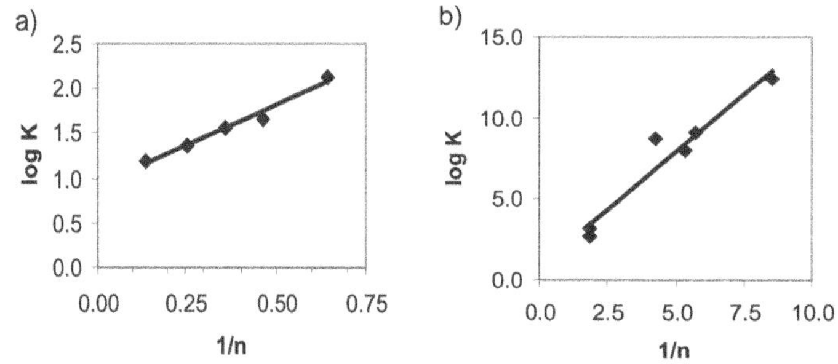

Figure 6-3. Log K and 1/n Correlations in a) Ultra-pure water and b) Ultra-pure Water Spiked with Effluent Organic Matter (TOC = 5 mg/L).

6.6 QSPR Toolbox

Table 6-5 lists currently the most viable QSPR components, which can be coupled into mass balance models to predict the removal of individual compounds through wastewater treatment trains. Table 6-5 lists the references for key QSPR models for various removal mechanisms that can be potentially be present during a given wastewater treatment train.

The following is guidance on how to use QSPR models for predicting the removal of target compounds:

1) Find the appropriate QSPR for the target compound:
   a. Does the target compound belong to the class of compounds used to develop the model?
   b. Do the water quality and operational parameters for the treatment process fall within the boundary conditions of the QSPR model?
2) Estimate (or measure) appropriate descriptor parameter(s).
   a. Calculate descriptor(s) with appropriate molecular modeling programs (Appendix E) by inputting the 2D and/or 3D molecular structure.
3) Calculate the treatment parameter by employing descriptor parameter(s) into the chosen QSPR model.
4) Predict treatment removal by incorporating the treatment parameter into the appropriate mass balance model.

It is advantageous and more practical to package the above steps in an overall predictive model, such as U.S. EPA's EPI Suite. The following is guidance on how to use computer programs with embedded QSPR/mass balance models for predicting the removal of target compounds:

1) Verify the target compound and water quality and treatment operational conditions fall within the boundary conditions of the predictive QSPR/mass balance model.
2) Input the 2D and/or 3D molecular structure.
3) Treatment removal is predicted.

From a utility viewpoint, a packaged QSPR predictive model is more practical. However, these types of models do not exist or are very limited for many of the removal mechanisms listed in Table 6-5, such as oxidation, activated carbon sorption, and membrane treatment processes.

Table 6-6. Important QSPR Components for Various Removal Mechanisms.

Removal Mechanism	Descriptor Parameters	Treatment Parameter	References
oxidation (ozone, hydroxyl radical)	Hammett constant, molecular orbital energies, polarity, hydrophobicity	degradation rate constant, percent removal	Gould (1987) Hoigne and Bader (1983) Hu et al. (2000) Tang (2004) Lei and Snyder (2007)
Hydrolysis	Hammett and steric constants, pKa	hydrolysis rate constant	Peijnenburg (1991)
Biodegradation	Hammett and steric constants, atomic charge, connectivity indices, hydrophobicity, Van der Waals radius	degradation rate constant, 5-day BOD	Okey and Stensel (1996) Raymond et al. (2001)
Volatilization	polarizability, connectivity indices, hydrogen bonding	Henry's law constant	Sabljic and Güsten (1989) Nirmalakhandan and Speece (1988)
soil sorption	water solubility, hydrophobicity, connectivity indexes	soil partitioning coefficient	Gawlik et al. (1997) Kahn et al. (2005) Sablijic (1987) Sabljic et al. (1995)
biosolids sorption	connectivity indices, hydrophobicity	biosolids partitioning coefficient	Dobbs et al. (1989) Hsieh and Mukherjee (2003)
activated carbon sorption	connectivity indices, water solubility, molar volume, molecular weight, Hammett constant, Taft's steric factor, polarizability, dipole moment, length of compound	carbon adsorbability, adsorption capacity, capacity reduction	Blum et al. (1994) Luehrs et al. (1996) Brasquet et al. (1997) Brasquet and Cloirec (1999) Crittenden et al. (1999) McElroy (2005) Li et al. (2005) Magnuson and Speth (2005)
physical separation, membranes	charge/polarity, molecular complexity, hydrophobicity, hydrogen bonding, surface area, length	percent rejection, orientation angle fitting parameter	Rodriquez et al. (2004) Bellona

6.7 Conclusions

QSPRs are viable empirical tools that can be coupled with mass balance models to quickly assess the fate and treatment of new and existing chemicals of concern within wastewater treatment systems. The removal of an organic chemical is dependent upon the compound's structure. There are a broad range and well-defined 2D- and 3D-based molecular descriptors that are easily calculable and potentially relevant to linking the structural features of a compound to its ability to be removed by a particular treatment mechanism. In some cases, 2D-based descriptors (e.g., connectivity) are sufficient in predicting a desired treatment parameter; however, 3D-based descriptors can be sometimes better predictors, such as volume, dipole moment, and polarizability for activated carbon adsorption of polar wastewater-derived contaminants. Existing QSPRs are limited, and their validity is in question for some removal mechanisms. Several important research gaps have been revealed in existing QSPR efforts:

1) Current 2D- and 3D-based QSPR models need to be comprehensively compared and evaluated in regards to their applicability to the removal wastewater-derived organic compounds for different wastewater treatment removal mechanisms. QSPRs are typically derived for narrow classes of compounds, experimentally matrix specific (e.g., soil, biomass, and activated-carbon type) and, for the most part, have been developed from simple laboratory-controlled experimental matrices. The compound removal is dependent upon structure and environmental conditions, and for some removal mechanisms the environmental conditions can be just as important in determining a compounds removal (e.g., biodegradation). Therefore, the accuracy of QSPRs needs to be evaluated for realistic environmental conditions (e.g., water matrix, compound concentration, and biomass, soil, activated carbon, membrane fouling types). For example, QSPRs used to establish wastewater treatment plant models, such as U.S. EPA's EPI Suite, should be further evaluated in regards to predicting the removal of wastewater-derived chemicals, especially polar EDCs, PhACs, PCPs and HHCs.

2) Uniformly measured treatment parameter data (e.g., degradation rate constants, partitioning coefficients) needs to be generated for wastewater-derived organic compounds, which can be used to evaluate existing QSPR models (where no literature values can be obtained), and, if needed, develop new QSPR models. The treatment data needs to be collected under environmental and operational conditions in which the organic compounds are expected to reside.

3) 2D- and/or 3D-based QSPR models need to be developed where no QSPRs exist for certain removal/transformation mechanisms (e.g., chlorination) or when existing QSPRs are not able to sufficiently predict the desired target compounds.

4) QSPRs need to be incorporated into mass balance models to form packaged QSPR predictive models that can be easily used and applied to certain wastewater treatment process (e.g., ozonation, AOPs, chlorination, GAC adsorption, membranes).

CHAPTER 7.0

CONCLUSIONS AND RECOMMENDATIONS

This study was tailored to address four primary objectives related to occurrence and fate of HPV household chemicals and chemicals in household commodities in wastewater systems:

1. To identify organic compounds from households and household commodities potentially present in municipal wastewater based on high-volume production, consumption pattern, and physicochemical properties.
2. To determine occurrence and fate of selected HPV target compounds through conventional and advanced wastewater treatment unit operations.
3. To compile a database on HPV chemicals and organic compounds in household commodities.
4. To identify and develop quantitative structure activity and property relationships that can be used to model the behavior of individual compounds through treatment processes.

Of the 26 household chemicals targeted in this study, 20 compounds were consistently detected in raw influents of full-scale wastewater treatment plants. Chemicals that are primarily used in products applied outdoors, such as pesticides (e.g., simazine, atrazine, trifluralin, and acriflavine), were generally not present in raw influent samples. The majority of compounds present in personal care and cleaning products generally appeared in all influent samples with concentrations of 2-phenoxyethanol (a preservative with various uses) and menthol (a fragrance with various uses) consistently exhibiting the highest concentrations of all compounds.

Occurrence in raw sewage and treated effluent among the six facilities showed very similar concentration ranges, supporting the studied HPV compounds are contributed by general consumers/households, rather than certain industries. The concept employed in this study for selecting HPV chemicals was confirmed. In general, concentrations of Tier 1 compounds in raw sewage were in the microgram-per-liter range, whereas Tier 2 compounds were present at several hundred nanograms-per-liter. In treated wastewater effluents, the Tier 1 compounds dibutyl phthalate, benzophenone, 2-phenoxytethanol, vanillin, and triclosan exhibited average effluent concentrations exceeding 200 ng/L with dibutyl phthalate exhibiting the highest average concentration of approximately 600 ng/L. Only three out of 13 Tier 2 compounds, 3-indolebutyric acid, DEET, and menthol, exhibited concentrations exceeding 100 ng/L in secondary treated effluents. Most of the compounds occurred at concentrations of less than 200 ng/L, which is comparable to the occurrence of pharmaceutical residues. The average concentration of LAS in raw sewage almost approached 6 mg/L and concentration was reduced in secondary treated effluents to approximately 4,000 ng/L, representing the highest observed concentrations of all compounds targeted in this study.

Daily mass loading variations of all Tier 1 compounds suggested a rather constant discharge of household chemicals to sewage with discharge directly coupled to water consumption and thus linked to consumer discharge. Daily mass loadings of Tier 2 compounds were significantly lower than Tier 1 compounds but also followed the general diurnal trend. Higher weekend loading values for some compounds (i.e., dibutyl phthalate, oxybenzone, triclocarban, camphor, and DEET) may indicate that many of these compounds have significant usage in household products. Conversely, lower weekend loading values observed for benzophenone and o-phenylphenol may indicate that those compounds also have significant uses other than household (i.e., industrial). Daily LAS mass loading exhibited a pattern following the common diurnal flow regime. This finding is expected since LAS is widely used as surfactants in many cleaning products and its use seems to be closely coupled with human activities and water consumption. Weekly sampling, however, revealed a significant higher loading of LAS during the weekend, which might be a consequence of more cleaning/laundry activities during the weekend.

Occurrence in the aqueous phase of sewage seems to depend not only upon production volume, but also compound properties. Of the three compounds produced at between 10 and 50 million pounds per year (BHT, dibutyl phthalate, and 2-phenoxyethanol), 2-phenoxyethanol is the most water soluble (log K_{ow} of 1.16) exhibiting the highest mass loading in raw sewage. BHT is the most hydrophobic compound (log K_{ow} = 5.1) and exhibited the lowest aqueous mass loading. A similar trend was observed for compounds produced between 1 and 10 million pounds per year, with vanillin being the most hydrophilic (log K_{ow} 1.21), exhibiting the highest mass loading and triclocarban, being the most hydrophobic (log K_{ow} 4.9) and exhibiting the lowest aqueous phase mass loading. Heidler et al. (2006) reported that for triclocarbon partitioning to filterable particulates in raw sewage averaged 76% (± 11%) indicating that, for hydrophobic compounds, a significant mass loading is bound to solids entering the plant. For compounds produced in smaller volume bins, no clear relationship between production volume and hydrophobicity was observed. Other HPV chemicals not addressed in this study, which have similar properties to compounds detected in raw wastewaters in this study, are also likely present in the raw water. An in-depth analysis (e.g., QSPR) of the properties of compound examined in this study will help in predicting how many other HPV chemicals will be present and perhaps at what concentrations in raw wastewaters.

The removal of HPV chemicals by biological treatment (activated sludge and MBR), disinfection (UV, chloramines, chlorine), and advanced oxidation (ozone and ozone/hydrogen peroxide) processes was evaluated. In general MBR and activated-sludge treatments removed the Tier 1 and 2 compounds similarly. All Tier 1 and 2 compounds that were detected were partially or completely removed (> 80%), indicating that biological treatment is a good treatment option for these compounds. For conventional disinfection processes chlorination represents the best process for partial removal of Tier 1 and 2 compounds and a second barrier. Chloramine and UV disinfection processes are not viable removal barriers. Ozonation and ozone/hydrogen peroxide treatments performed similarly and are additional barrier options where higher effluent quality is desired. However, they do not represent a 100% barrier for all compounds (e.g., vanillin).

QSPRs are viable empirical tools that can be coupled with mass balance models to quickly assess the fate and treatment of new and existing chemicals of concern within wastewater treatment systems. The removal of an organic chemical is dependent upon the compound's structure. There is a broad range of well-defined molecular descriptors (e.g., 3D-based descriptors) that are easily calculable and potentially relevant to linking the structural features of a compound to its ability to be removed by a particular treatment mechanism. Several existing QSPRs that are relevant to significant wastewater treatment removal routes, reactivity, phase partitioning, and physical separation have been identified as viable tools that can be applied to the prediction of the removal of individual compounds through wastewater treatment trains. However, there is a need to evaluate QSPRs for their applicability to a wide range of organic compounds and realistic environmental conditions. Also, QSPRs need to be evaluated in regards to which QSPRs best describe chemically and/or physically an organic compound's potential to be removed by a particular removal mechanism. For some removal mechanisms there is a lack of uniformly measured treatment data (e.g., biodegradation rate constants). Therefore, there is a need to generate fate parameter data, particularly biodegradation and chlorination removal rates are lacking, for the evaluation of existing and development of new applicable QSPR models. Finally, QSPRs need to be incorporated into appropriate mass balance models and these mass balance models (e.g., fugacity models) need to be validated by full-scale treatment data.

APPENDIX A

FACILITY DESCRIPTIONS

Facility 1:

Treatment

In 2003, Facility 1 had an operational capacity of 62.5 million mgd but was operated at 57.5 mgd. It receives approximately 94% residential and 6% industrial sewage, respectively. The service area has no combined sewers; therefore, no stormwater run-off is treated. Drinking water supply in the service area of the facility is a blend of surface and groundwater.

Treatment at Facility 1 consists of primary clarification (ferric chloride added prior to inlet), activated sludge treatment with nitrification and denitrification, secondary clarification, alum addition for coagulation and pre-filter disinfection using chloramination, tertiary dual-media filtration (anthracite, sand), and chlorine contact basin. Sulfur dioxide is used after chlorine contact to dechlorinate prior to discharge. The process schematic of Facility 1 is illustrated in Figure A-1.

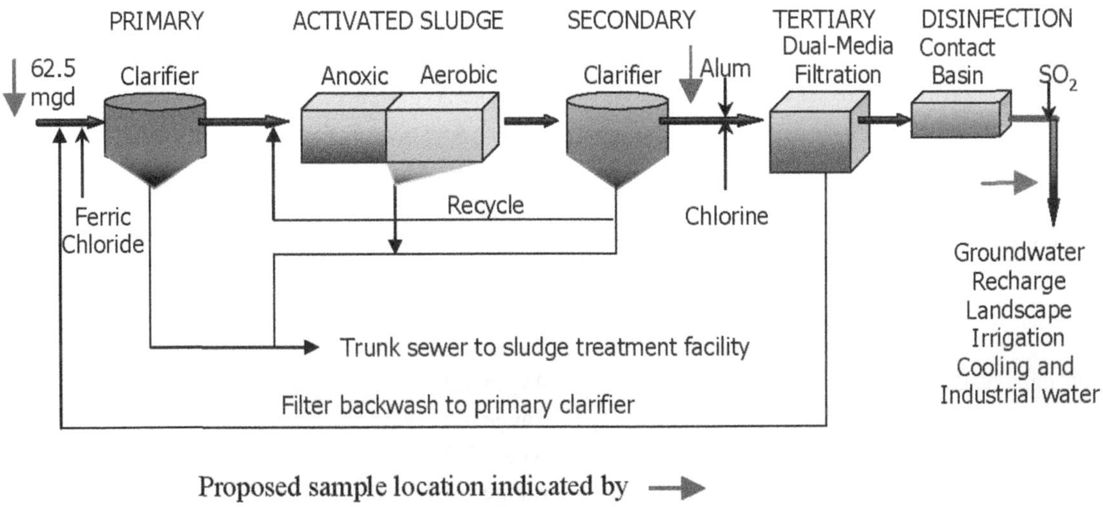

Figure A-1. Process Schematic of Facility 1.

The hydraulic retention time during primary treatment is approximately 1.65 hours at 57.5 mgd. The average mixed liquor aeration time varies between 2.5 and 3 hours and the solids retention time varied between 13 and 48 days. The design filtration time in the dual-media filters was 4 gallons/day/ft^2. Excess sludge is discharged to a trunk sewer and treated at a centralized treatment facility.

Water Quality

Key operational parameters and water quality of secondary treated wastewater at Facility 1 are summarized in Table A-1, which presents monthly average, minimum, and maximum values for the year 2003. The effluent quality meets California Title 22 standards for recycled water.

Table A-1. Facility 1 Key Operational Parameters and Treated Effluent Water Quality.

Key Operational Parameters		
Parameter	**Unit**	**Concentration**
pH	-	NA
BOD loading	kg/day	Avg = 18,130 Range = 14,963 - 21,191
BOD concentration	mg/L	Avg = 207 Range = 130 - 367
COD concentration	mg/L	Avg = 358 Range = 237 - 456
HRT primary	hours	1.65
Mixed liquor total suspended solids	mg/L	Avg = 3,418 Range = 2,036 - 4,303
Mixed liquor aeration time	hours	Avg = 2.7
SRT	days	Range = 15-48
Treated Effluent Parameters		
Parameter	**Unit**	**Concentration**
pH	-	Avg = 6.89 Range = 6.3 – 7.7
Temperature	°C	Summer Avg = 28.7 Winter Avg = 24.4
Conductivity	µS/cm	Avg = 1070 Range = 810 – 1,300
BOD concentration	mg/L	Avg = 3.6 Range = 2 – 9
COD concentration	mg/L	Avg = 22.6 Range = 12 – 40
TOC (no DOC data available)	mg/L	Avg = 7.2 Range = 5 – 9.2
NH_4-N	mg/L	Avg = 3.9 Range = 1.1 – 10.2
NO_3^--N	mg/L	Avg = 3.8 Range = 2.36 – 4.75
PO_4^{3-}-P (reported semi-annually)	mg/L	Range = 0.8 – 5

Sampling Strategy

Two, 24-hour composite samples were collected from Facility 1 to evaluate the presence and overall removal of the selected target compounds. These samples included raw sewage at the inlet of the primary clarifier (prior to ferric chloride addition), and final effluent after dechlorination. In addition, three 4-hour composite samples were collected to examine the removal efficiencies of the target compounds during individual unit operations. These samples included raw sewage, secondary treated effluent at the discharge of the secondary clarifier prior to chlorination and alum addition, and final effluent. The 4-hour composite samples at each unit operation generally considered the hydraulic retention time of each unit. Sampling information is summarized in Table A-2.

Table A-2. Facility 1 – Sampling Information.

Sample Location	Samples (type and date)	Samples (type and date)	Parameters	Analytical Responsibility	Container
Raw sewage prior to primary clarifier	12-hr composite April 2005	4-hr composite Time: 0 October 2005	pH, T, conductivity, NH_4-N, NO_3-N, PO_4-P	Facility 1	NA (on-site)
			pH, T, conductivity, DOC, UV	CSM	1 x 250 mL glass amber bottle
			pH, T GC-MS; LC/MS-MS	SNWA	2 x 1 L glass amber bottle
Secondary effluent at weir of secondary clarifier	NA	4-hr composite Time: 0 + 5.5 hrs October 2005	pH, T, conductivity, NH_4-N, NO_3-N, PO_4-P, MLTSS	Facility 1	NA (on-site)
			pH, T, conductivity, DOC, UV	CSM	1 x 250 mL glass amber bottle
			pH, T GC-MS; LC/MS-MS	SNWA	2 x 1 L glass amber bottle
Tertiary effluent after dechlorination	12-hr composite April 2005	4-hr composite Time: 0 + 6 hrs October 2005	pH, T, conductivity, NH_4-N, NO_3-N, PO_4-P	Facility 1	NA (on-site)
			pH, T, conductivity, DOC, UV	CSM	1 x 250 mL glass amber bottle
			pH, T GC-MS; LC/MS-MS	SNWA	2 x 1 L glass amber bottle

Facility 2 – Full-Scale Water Reclamation Plant and On-Site Zenon Membrane Bioreactor Pilot-Plant:

Treatment

Facility 2 is designed to treat 15 mgd but generally operates at 10 mgd. It receives approximately 90% residential sewage and 10% industrial discharge. The service area has no combined sewers; therefore, no stormwater run-off is treated. Drinking water supply in the service area of the facility is a blend of surface and groundwater.

Treatment at Facility 2 consists of primary clarification, activated sludge treatment with nitrification/denitrification through a sequence of anoxic and aerobic zones, secondary clarification, alum addition for coagulation prior to filtration, disinfection using sodium hypochlorite (pre-filter chlorination is used), tertiary dual-media filtration (anthracite, sand), and chlorine contact basin. Ammonia is added prior to chlorination to achieve chloramination. Sodium bisulfite is used after chlorine contact to dechlorinate prior to discharge.

The on-site membrane bioreactor pilot-plant currently receives approximately 25 gpm of primary effluent from Facility 2 but has an operational capacity of 25 to 50 gpm. Pilot plant treatment includes secondary activated sludge treatment with nitrification and denitrification, followed by tertiary membrane filtration using two membrane cassettes, each containing 10 Zenon ZeeWeed® 500c hollow fiber membrane elements with a nominal pore size of 0.04 micrometers. The process schematic of the full-scale facility and the membrane bioreactor pilot-plant are illustrated in Figures A-2 and A-3, respectively.

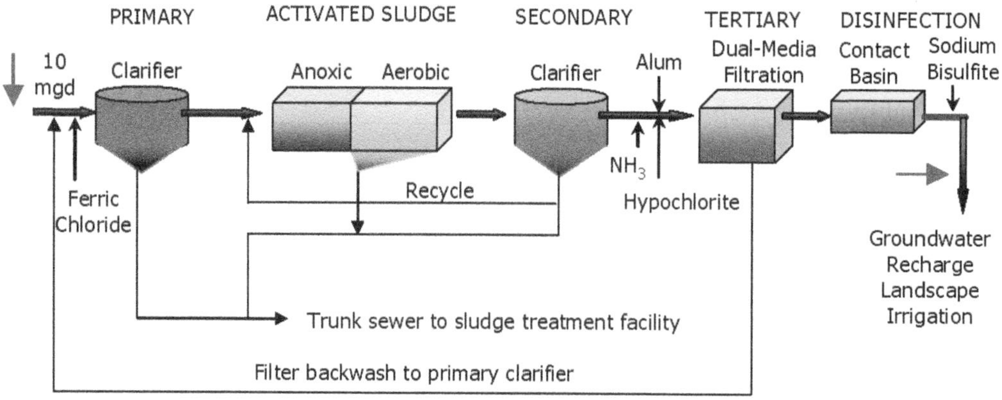

Proposed sample location indicated by →

Figure A-2. Facility 2 Process Schematic.

The hydraulic retention time for Facility 2's primary treatment is approximately 3.2 hr at 10 mgd. The average mixed liquor aeration time is 3.2 hr, and the solids retention time is approximately 19 days. The average detention time in the dual-media filters is 30 minutes. The contact time in the chlorination contact basin is approximately 90 minutes. For fiscal year 2001-2002, 98.3% of the effluent produced at this facility was actively reused for groundwater recharge and irrigation purposes. The great majority (99.2%) of the reclaimed water was used to recharge the local groundwater aquifer. The remaining 0.8% of reclaimed water was used for ornamental plant irrigation at a local nursery. The facility does not employ on-site solids treatment. Excess sludge is discharged to a trunk sewer and treated at a centralized treatment facility.

The hydraulic retention time for the pilot-plant during primary treatment is approximately 3 hours. The average mixed liquor aeration time varies between 5 and 8 hours and the solids retention time varies between 10 and 15 days. The membrane filter has a flux capacity of 9 to 18 gal/ft^2·day. With a total membrane surface area of 4,730 ft^2, the current flux of 7.6 gal/ft^2·day is currently below operating capacity.

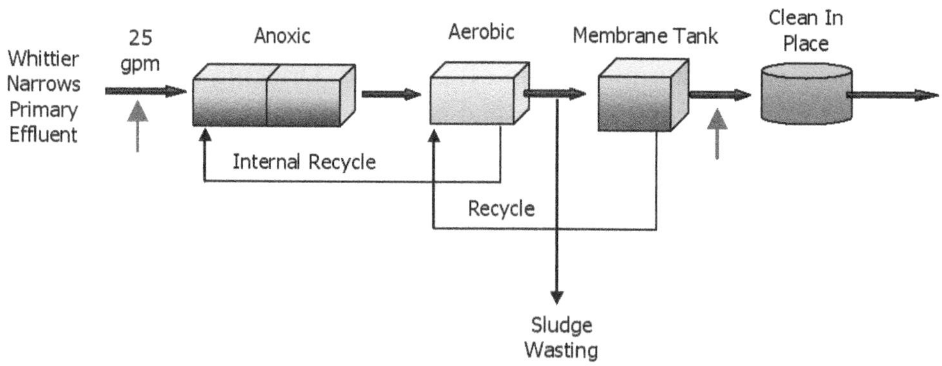

Proposed sample location indicated by →

Figure A-3. Facility 2 On-Site Membrane Bioreactor Pilot-Plant Process Schematic.

Water Quality

Key operational parameters and water quality of secondary treated wastewater at Facility 2 are summarized in Table A-3, which presents monthly average, minimum, and maximum values for the time period 1997-2002. The effluent quality meets California Title 22 standards for recycled water.

Key operational parameters and water quality of secondary treated wastewater at the pilot-plant are summarized in Table A-4, which also includes monthly average, minimum, and maximum values for the year 2003.

Sampling Strategy

Because the on-site membrane bioreactor pilot-plant receives primary effluent from Facility 2, the pilot-plant was only considered for unit operations performance sampling and was not been included during occurrence monitoring, while Facility 2 was only considered for occurrence monitoring sampling.

During occurrence monitoring, two 24-hour composite samples were collected from Facility 2 to evaluate the presence and overall removal of the selected target compounds. These samples included raw sewage and final effluent after dechlorination. During performance monitoring, samples for the Facility 2 membrane bioreactor pilot-plant included two 4-hour composite samples collected from the Facility 2 primary effluent at the inlet of the pilot-plant anoxic tank Number 1, and from the discharge of the membrane tank. Sampling information for Facility 2 and the pilot plant are summarized in Tables A-5 and A-6, respectively.

Table A-3. Facility 2 Key Operational Parameters and Treated Effluent Water Quality.

Key Operational Parameters		
Parameter	**Unit**	**Concentration**
pH	-	NA
BOD loading	kg/day	Avg = 4,990 Range = 2,812 – 6,803
BOD concentration	mg/L	NA
COD concentration	mg/L	NA
HRT primary	hours	3.2
HRT secondary	hours	3.2
Mixed liquor total suspended solids	mg/L	Avg = 3,400 Range = 1,760 - 3,500
Mixed liquor aeration time	hours	NA
SRT	days	19
Treated Effluent Parameters		
Parameter	**Unit**	**Concentration**
pH	-	Avg = 7.3 Range = 7.0 - 7.4
Temperature	°C	NA
Conductivity	µS/cm	Avg = 940 Range = 890 - 1,120
BOD concentration	mg/L	Avg = 2 Range = 1.1 – 4
COD concentration	mg/L	Avg = 23 Range = 18 – 31
DOC (no TOC data available)	mg/L	Avg = 7.3 Range = 6.3 - 12.8
NH_4-N	mg/L	Avg = 1.2 Range = 0.1 - 3.8
NO_3^--N	mg/L	Avg = 6.2 Range = 4.0 - 7.8
PO_4^{3-}-P	mg/L	Avg = 2.6 Range = 0.7 - 4.7

Table A-4. Facility 2 On-Site Membrane Bioreactor Pilot-Plant - Key Operational Parameters and Treated Effluent Water Quality.

Key Operational Parameters		
Parameter	**Unit**	**Concentration**
pH	-	Avg = 7.6 Range = 7.5 – 7.8
BOD loading	kg/day	Avg = 28 Range = 23 - 35
BOD concentration	mg/L	Avg = 165 Range = 120 - 230
COD concentration	mg/L	Avg = 323 Range = 240 - 395
HRT primary	hours	3
Mixed liquor total suspended solids	mg/L	Avg = 7,950 Range = 7,500 – 8,500
Mixed liquor aeration time	hours	Avg = 6.7
SRT	days	Range = 10 -15
Treated Effluent Parameters		
Parameter	**Unit**	**Concentration**
pH	-	NA
Temperature	°C	Summer Avg = 26.1 Winter Avg = 23.6
Conductivity	µS/cm	NA
BOD concentration	mg/L	Avg = < 3 Range = < 3
COD concentration	mg/L	Avg = 13 Range = 10 - 18
TOC/DOC	mg/L	NA
NH_4-N	mg/L	Range = < 1 – 2.2
NO_3^--N	mg/L	Avg = 6.2 Range = 4.4 – 11.1
PO_4^{3-}-P	mg/L	NA

Table A-5. Facility 2 Sampling Information.

Sample Location	Samples (type, date)	Parameters	Analytic Responsibility	Container
Raw sewage prior to primary clarifier	24-hr composite	pH, T, conductivity, NH_4-N, NO_3-N, PO_4-P	Facility 2	NA (on-site)
	May 2005	pH, T, conductivity, DOC, UV	CSM	1 x 250 mL glass amber bottle
		pH, T GC-MS; LC/MS-MS	SNWA	2 x 1 L glass amber bottle
Final treated effluent after dechlorination	24-hr composite	pH, T, conductivity, NH_4-N, NO_3-N, PO_4-P	Facility 2	NA (on-site)
	May 2005	pH, T, conductivity, DOC, UV	CSM	1 x 250 mL glass amber bottle
		pH, T GC-MS; LC/MS-MS	SNWA	2 x 1 L glass amber bottle

Table A-6. Facility 2 Membrane Bioreactor Pilot-Plant – Sampling Information.

Sample Location	Samples (type and date)	Parameters	Analytic Responsibility	Container
Facility 2 primary influent at anoxic tank No. 1 inlet	4-hr composite Time: 0 hrs September 2005	pH, T, conductivity, NH_4-N, NO_3-N, PO_4-P	Facility 2	NA (on-site)
		pH, T, conductivity, DOC, UV	CSM	1 x 250 mL glass amber bottle
		pH, T GC-MS; LC/MS-MS	SNWA	2 x 1 L glass amber bottle
Treated effluent at membrane tank discharge	4-hr composite Time: 0 + 5 hrs September 2005	pH, T, conductivity, NH_4-N, NO_3-N, PO_4-P	Facility 2	NA (on-site)
		pH, T, conductivity, DOC, UV	CSM	1 x 250 mL glass amber bottle
		pH, T GC-MS; LC/MS-MS	SNWA	2 x 1 L glass amber bottle

Facility 3:

Treatment

Facility 3 comprises individual north and south treatment processes. In 2003, both the north and south process had an operational capacity of 120 mgd, but because of drought conditions at the time, were operated at 99.3 mgd and 34. mgd, respectively. Facility 3 receives approximately 93% of its wastewater as residential influent and 7% as industrial influent. Stormwater run-off is not treated within the service area. Drinking water supply in the service area of the facility is primarily surface water (78%) but does include groundwater (22%).

Treatment at Facility 3 consists of grit removal and primary clarification, activated sludge treatment with nitrification and denitrification, secondary clarification, and chlorine contact basin. Sulfur dioxide is used after chlorine contact to dechlorinate prior to discharge. The process schematic of Facility 3 is illustrated in Figure A-4.

The hydraulic retention time during primary treatment varied between 1.6 and 2.1 hours for the north treatment process and between 2.4 and 3.9 hours for the south treatment process. The average mixed liquor aeration time varied between 5.2 and 7.1 hours for the north process and between 1.7 and 2.4 hours for the south process. Solids retention time varied between 4 and 8.4 days for the north process and between 1.2 and 2.3 days for the south process. Information on the contact time in the chlorination contact basin was not provided. Solids are treated on site through a combination of anaerobic digestion and centrifugation with subsequent land application for beneficial reuse (95%) or composting for home use (5%).

Water Quality

Key operational parameters and water quality of secondary treated wastewater at Facility 3 are summarized in Table A-7, which presents monthly average, minimum, and maximum values for the year 2003.

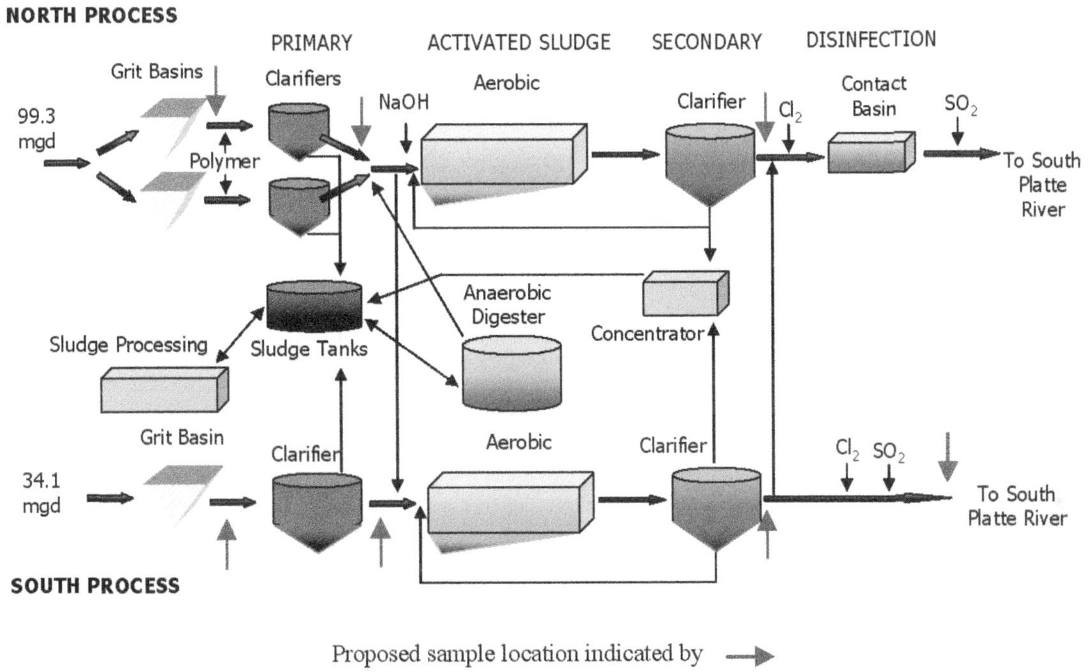

Figure A-4. Facility 3 Process Schematic.

Sampling Strategy

During occurrence monitoring, two 24-hour composite samples were collected from both the North and the South treatment process (four samples total) of Facility 3 to evaluate the presence and overall removal of the selected target compounds. These samples include raw sewage at the discharge of both the north and south grit basins (one each), and secondary treated effluent prior to disinfection in the north process, as well as secondary treated effluent after disinfection and chlorine removal in the south process. During performance monitoring, five 4-hour composite samples will be collected to examine the removal efficiencies of the target compounds during individual unit operations. These samples will include primary effluent at the discharge of the primary clarifiers for both processes, secondary treated effluent at the discharge of the secondary clarifiers for both processes (and prior to disinfection), and tertiary effluent after disinfection and chlorine removal in the south process. The 4-hour composite samples at each unit operation will generally consider the hydraulic retention time of each unit. Sampling information for both the north and south treatment process are summarized below in Tables A-8 and A-9, respectively.

Table A-7. Facility 3 Key Operational Parameters and Treated Effluent Water Quality.

Key Operational Parameters			
Parameter	**Units**	**North Process Concentration**	**South Process Concentration**
pH	-	Not Determined	Not Determined
BOD loading	kg/day	Avg = 103,154	Avg = 36,295
		Range = 80,739 – 136,078	Range = 22,680 – 49,895
BOD concentration	mg/L	Avg = 216	Avg = 193
		Range = 182 - 268	Range = 173 - 225
COD concentration	mg/L	Not Determined	Not Determined
HRT primary	Hours	Range = 1.6 – 2.1	Range = 2.4 – 3.9
Mixed liquor total suspended solids	mg/L	Avg = 3,004	Avg = 2,584
		Range = 2,170 – 3,620	Range = 1,820 – 3,270
Mixed liquor aeration time	Hours	Avg = 6.7	Avg = 6.7
		Range = 5.2 – 7.1	Range = 1.7 – 2.4
SRT	days	Avg = 6.1	Avg = 1.7
		Range = 4 – 8.4	Range = 1.2 – 2.3
Treated Effluent Parameters			
Parameter	**Units**	**North Process Concentration**	**South Process Concentration**
pH	-	Avg = 6.68	Avg = 6.63
		Range = 6.5 – 7	Range = 6.45 – 6.9
Temperature	°C	Avg = 18.5	Avg = 18.2
		Range = 15.4 – 22.9	Range = 14.8 - 22
Conductivity	µS/cm	Avg = 857	Avg = 951
		Range = 760 – 984	Range = 874 – 1,066
CBOD concentration (vs. BOD)	mg/L	Avg = 6.5	Avg = 10.4
		Range = 2 – 20	Range = 3 - 36
COD concentration	mg/L	Not Determined	Not Determined
DOC (TOC not determined)	mg/L	Avg = 10.8	Avg = 13.6
		Range = 10 – 12	Range = 9 - 16
NH_4-N	mg/L	Avg = 3.3	Avg = 17.6
		Range = 1.4 – 10.7	Range = 11.2 – 21.8
NO_3^--N (value includes NO_2)	mg/L	Avg = 8.52	Avg = 0.97
		Range = 1.45 – 13.3	Range = 0.0 – 10.9
PO_4^{3-}-P	mg/L	Avg = 2.18	Avg = 1.99
		Range = 0.99 – 2.82	Range = 0.83 – 2.66

Table A-8. Facility 3 North Treatment Process – Sampling Information.

Sample Location	Samples (type and date)	Samples (type and date)	Parameters	Analytic Responsibility	Container
Raw sewage at grit basin discharge	24-hr composite March 2005	NA	pH, T, conductivity, NH_4-N, NO_3-N, PO_4-P	Facility 3	NA (on-site)
			pH, T, conductivity, DOC, UV	CSM	1 x 250 mL glass amber bottle
			pH, T GC-MS; LC/MS-MS	SNWA	2 x 1 L glass amber bottle
Primary treated effluent at primary clarifier discharge	NA	4-hr composite Time: 0 hrs September 2005	pH, T, conductivity, NH_4-N, NO_3-N, PO_4-P, MLTSS	Facility 3	NA (on-site)
			pH, T, conductivity, DOC, UV	CSM	1 x 250 mL glass amber bottle
			pH, T GC-MS; LC/MS-MS	SNWA	2 x 1 L glass amber bottle
Secondary treated effluent at secondary clarifier discharge	24-hr composite March 2005	4-hr composite Time: 0 + 7 hrs September 2005	pH, T, conductivity, NH_4-N, NO_3-N, PO_4-P, MLTSS	Facility 3	NA (on-site)
			pH, T, conductivity, DOC, UV	CSM	1 x 250 mL glass amber bottle
			pH, T GC-MS; LC/MS-MS	SNWA	2 x 1 L glass amber bottle

Table A-9. Facility 3 South Treatment Process – Task 6 and Task 7 Sampling Information.

Sample Location	Samples (type and date)	Samples (type and date)	Parameters	Analytic Responsibility	Container
Raw sewage at grit basin discharge	24-hr composite March 2005	NA	pH, T, conductivity, NH_4-N, NO_3-N, PO_4-P	Facility 3	NA (on-site)
			pH, T, conductivity, DOC, UV	CSM	1 x 250 mL glass amber bottle
			pH, T GC-MS; LC/MS-MS	SNWA	2 x 1 L glass amber bottle
Primary treated effluent at primary clarifier discharge	NA	4-hr composite Time: 0 hrs December 2005	pH, T, conductivity, NH_4-N, NO_3-N, PO_4-P, MLTSS	Facility 3	NA (on-site)
			pH, T, conductivity, DOC, UV	CSM	1 x 250 mL glass amber bottle
			pH, T GC-MS; LC/MS-MS	SNWA	2 x 1 L glass amber bottle
Secondary treated effluent at secondary clarifier discharge	NA	4-hr composite Time: 0 + 4 hrs December 2005	pH, T, conductivity, NH_4-N, NO_3-N, PO_4-P, MLTSS	Facility 3	NA (on-site)
			pH, T, conductivity, DOC, UV	CSM	1 x 250 mL glass amber bottle
			pH, T GC-MS; LC/MS-MS	SNWA	2 x 1 L glass amber bottle
Tertiary effluent after disinfection and chlorine removal	24-hr composite March 2005	4-hr composite Time: 0 + 4.5 hrs December 2005	pH, T, conductivity, NH_4-N, NO_3-N, PO_4-P	Facility 3	NA (on-site)
			pH, T, conductivity, DOC, UV	CSM	1 x 250 mL glass amber bottle
			pH, T GC-MS; LC/MS-MS	SNWA	2 x 1 L glass amber bottle

Facility 4:

Treatment

Facility 4 was constructed in 2004 with an operational design capacity of 30 mgd but is operated at 2.5 to 18 mgd. Facility 4 receives secondary clarifier effluent from Facility 3, which receives approximately 93% of its wastewater as residential influent and 7% as industrial influent. As stated previously, stormwater run-off is not treated within the service area and drinking water supply in the service area is primarily surface water (78%) but does include groundwater (22%).

Treatment at Facility 4 consists of nitrification using biological aerated filtration (BAF), coagulation/flocculation prior to secondary clarification, tertiary single-media filtration (anthracite), and chlorine contact basin. The process schematic of Facility 4 is illustrated in Figure A-5. No information was provided on hydraulic retention time during biological aerated filtration; however, the average detention time in the single-media filters varies between 2.7 and 7.2 minutes. At the maximum operational capacity of 30 mgd, the estimated contact time in the chlorination contact basin is approximately 30 minutes. Solids are treated on site through a combination of anaerobic digestion and centrifugation.

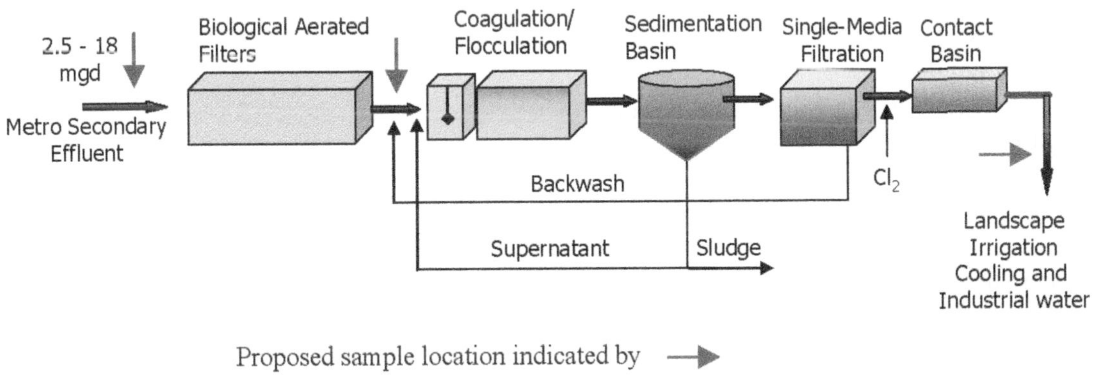

Figure A-5. Facility 4 Process Schematic.

Water Quality

Key operational parameters and water quality of secondary treated wastewater at Facility 4 are summarized in Table A-10, which presents monthly average, minimum, and maximum values for the year 2004.

Table A-10. Facility 4 Key Operational Parameters and Treated Effluent Water Quality.

Key Operational Parameters		
Parameter	**Unit**	**Concentration**
pH	-	Avg = 6.9 Range = 6.52 – 8.17
BOD loading	kg/day	NA
BOD concentration	mg/L	Range = < 3 – 47
COD concentration	mg/L	NA
HRT primary	Hours	NA
Mixed liquor total suspended solids	mg/L	NA
Mixed liquor aeration time	Hours	NA
SRT	Days	NA
Treated Effluent Parameters		
Parameter	**Unit**	**Concentration**
pH	-	Avg = 7.2 Range = 6.48 – 8.25
Temperature	°C	Avg = 19.4 Range = 8 – 25
Conductivity	µS/cm	Avg = 950 Range = 810 – 1,240
BOD concentration	mg/L	Range = < 3 – 15.8
COD concentration	mg/L	NA
TOC (single data value – no DOC data reported)	mg/L	7.1
NH_4-N	mg/L	Range = 0.0 – 0.162
NO_3^--N	mg/L	Avg = 11.6 Range = 6.6 – 14.8
PO_4^{3-}-P	mg/L	Avg = 0.16 Range = 0.0 – 0.67

Sampling Strategy

Because Facility 4 receives secondary treated effluent from Facility 3, Facility 4 was only considered for performance monitoring sampling. During monitoring sampling, three 4-hour composite samples were collected to examine the removal efficiencies of the target compounds during individual unit operations. These samples included secondary effluent at the inlet of the BAF, effluent at the discharge of the BAF, and tertiary effluent after disinfection. The 4-hour composite samples at each unit operation generally considered the hydraulic retention time of each unit. Sampling information is summarized below in Table A-11.

Table A-11. Facility 4 Sampling Information.

Sample Location	Samples (type and date)	Parameters	Analytic Responsibility	Container
Secondary treated effluent at BAF inlet	4-hr composite Time: 0	pH, T, conductivity, NH_4-N, NO_3-N, PO_4-P	Facility 4	NA (on-site)
	September 2005	pH, T, conductivity, DOC, UV	CSM	1 x 250 mL glass amber bottle
		pH, T GC-MS; LC/MS-MS	SNWA	2 x 1 L glass amber bottle
Treated effluent at BAF discharge	4-hr composite Time: 0 + 2 hrs	pH, T, conductivity, NH_4-N, NO_3-N, PO_4-P, MLTSS	Facility 4	NA (on-site)
	September 2005	pH, T, conductivity, DOC, UV	CSM	1 x 250 mL glass amber bottle
		pH, T GC-MS; LC/MS-MS	SNWA	2 x 1 L glass amber bottle
Tertiary effluent after disinfection	4-hr composite Time: 0 + 2.5 hrs	pH, T, conductivity, NH_4-N, NO_3-N, PO_4-P	Facility 4	NA (on-site)
		pH, T, conductivity, DOC, UV	CSM	1 x 250 mL glass amber bottle
	September 2005	pH, T GC-MS; LC/MS-MS	SNWA	2 x 1 L glass amber bottle

Facility 5:

Treatment

In 2003, Facility 5 had an operational capacity of 110 mgd serving a population of approximately 900,000, but was operated at approximately 88.5 mgd. Industry accounts for only a small percentage of wastewater influent (percentage not available) and includes primarily casino/resort industry. No stormwater sewers are serviced. Drinking water supplies within the District is provided by groundwater, although no wells are located within the facility service area.

Treatment at Facility 5 consists of primary clarification, activated sludge treatment with nitrification and denitrification, secondary clarification, tertiary dual-media filtration (anthracite, sand), and ultraviolet (UV) disinfection. The process schematic of Facility 5 is illustrated in Figure A-6. The hydraulic retention time during primary treatment averaged from 1.5 to 2 hours at 88.5 mgd. The average mixed liquor aeration time varied between 2.9 and 4.8 hours and the solids retention time varied between 3.9 and 15 days. The average detention time in the dual-media filters was approximately 17 minutes. The facility does not employ on-site solids treatment. Excess sludge is dewatered and landfilled.

Water Quality

Key operational parameters and water quality of secondary treated wastewater at Facility 5 are summarized in Table A-12, which presents monthly average, minimum, and maximum values for the year 2003.

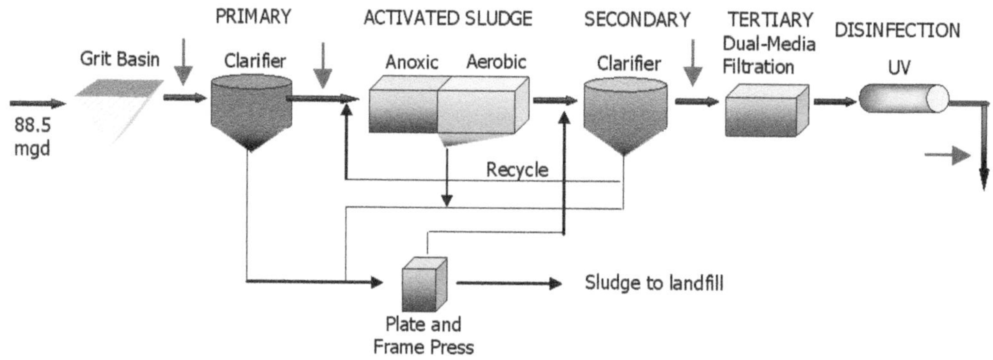

Proposed sample location indicated by →

Figure A-6. Process Schematic of Facility 5.

Table A-12. Facility 5 Key Operational Parameters and Treated Effluent Water Quality.

Key Operational Parameters		
Parameter	**Unit**	**Concentration**
pH	-	NA
BOD loading	kg/day	Avg = 52,550 Range = 37,249 - 69,057
BOD concentration	mg/L	Avg = 149 Range = 78 – 314
COD concentration	mg/L	NA
HRT primary	Hours	1.5 – 2
Mixed liquor total suspended solids	mg/L	Avg = 3,006 Range = 1,824 - 4,025
Mixed liquor aeration time	Hours	Avg = 3.4 Range = 2.9 – 4.8
SRT	Days	Avg = 6.17 Range = 3.9 – 15
Treated Effluent Parameters		
Parameter	**Unit**	**Concentration**
pH	-	Avg = 7.01 Range = 6.5 – 7.65
Temperature	°C	Avg = 23.7 Range = 17.3 – 29.5
Conductivity	µS/cm	NA
BOD concentration	mg/L	Avg = 2.1 Range = 1 – 7
COD concentration	mg/L	NA
TOC/DOC	mg/L	NA
NH_4-N	mg/L	Avg = 0.27 Range = 0.02 – 2.69
NO_3^--N	mg/L	Avg = 15.02 Range = 11.52 – 25.88
PO_4^{3-}-P (reported semi-annually)	mg/L	Avg = 0.2 Range = 0.04 – 1.0

Sampling Strategy

During occurrence monitoring, two 24-hour composite samples were collected from Facility 5 to evaluate the presence and overall removal of the selected target compounds. These samples included raw sewage at the discharge of the grit basin, and tertiary effluent after UV disinfection. During performance monitoring, four 4-hour composite samples were collected to examine the removal efficiencies of the target compounds during individual unit operations. These samples included raw sewage, primary effluent at the discharge of the primary clarifier, secondary effluent at the discharge of the secondary clarifier, and tertiary effluent after disinfection. The 4-hour composite samples at each unit operation generally considered the hydraulic retention time of each unit. Sampling information is summarized below in Table A-13.

Table A-13. Facility 5 Sampling Information.

Sample Location	Samples (type and date)	Samples (type and date)	Parameters	Analytic Responsibility	Container
Raw sewage at grit basin discharge	24-hr composite June 2005	4-hr composite Time: 0 December 2005	pH, T, conductivity, NH_4-N, NO_3-N, PO_4-P	Facility 5	NA (on-site)
			pH, T, conductivity, DOC, UV	CSM	1 x 250 mL glass amber bottle
			pH, T GC-MS; LC/MS-MS	SNWA	2 x 1 L glass amber bottle
Primary effluent at primary clarifier discharge	NA	4-hr composite Time: 0 + 1.5 hrs December 2005	pH, T, conductivity, NH_4-N, NO_3-N, PO_4-P, MLTSS	Facility 5	NA (on-site)
			pH, T, conductivity, DOC, UV	CSM	1 x 250 mL glass amber bottle
			pH, T GC-MS; LC/MS-MS	SNWA	2 x 1 L glass amber bottle
Secondary treated effluent at secondary clarifier discharge	NA	4-hr composite Time: 0 + 4.5 hrs December 2005	pH, T, conductivity, NH_4-N, NO_3-N, PO_4-P, MLTSS	Facility 5	NA (on-site)
			pH, T, conductivity, DOC, UV	CSM	1 x 250 mL glass amber bottle
			pH, T GC-MS; LC/MS-MS	SNWA	2 x 1 L glass amber bottle
Tertiary effluent after disinfection	24-hr composite June 2005	4-hr composite Time: 0 + 6.5 hrs December 2005	pH, T, conductivity, NH_4-N, NO_3-N, PO_4-P	Facility 5	NA (on-site)
			pH, T, conductivity, DOC, UV	CSM	1 x 250 mL glass amber bottle
			pH, T GC-MS; LC/MS-MS	SNWA	2 x 1 L glass amber bottle

Facility 6:

Treatment

In 2003, Facility 6 had a design capacity of 24 mgd serving approximately 235,000 people, but was operated at an average capacity of 21 mgd. Facility 6 receives a combination of industrial (various) and domestic wastewater (percentages unknown). Stormwater sewers are not serviced by the facility. Surface water from the Colorado River provides drinking water to the service area.

Treatment at Facility 6 consists of grit removal, activated sludge treatment with nitrification and denitrification, secondary clarification, addition of alum and polymer for phosphorous removal prior to filtration, tertiary single-media filtration (sand), and chlorine disinfection (sodium hypochlorite). The process schematic for Facility 6 is illustrated below in Figure A-7. The average mixed liquor aeration time varied between 10.5 and 11.5 hours and the solids retention time varied between 4 and 16 days. The average detention time in the single-media filters was approximately 12 minutes. Sludge is dewatered on site and then landfilled.

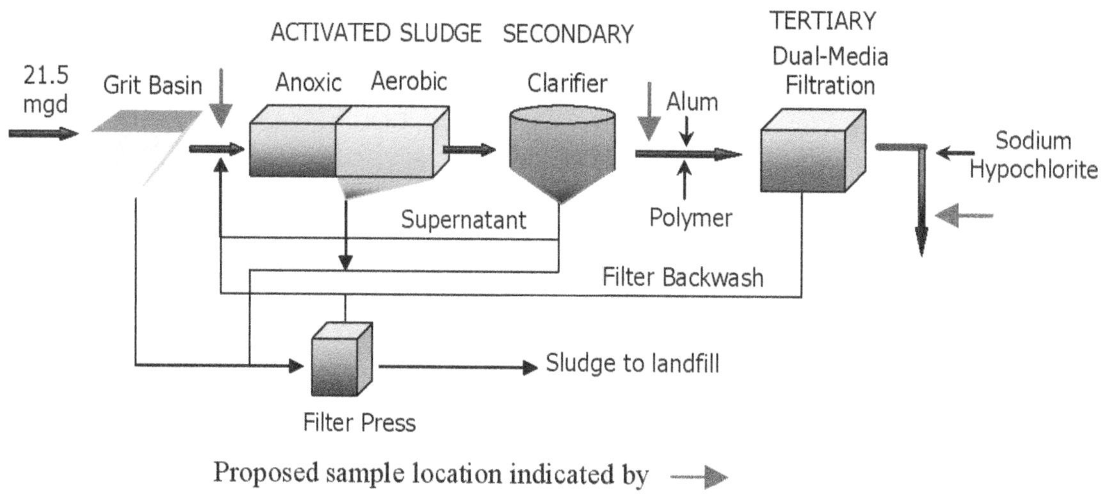

Figure A-7. Process Schematic of Facility 6.

Water Quality

Key operational parameters and water quality of treated effluent at Facility 6 are summarized in Table A-14, which presents monthly average, minimum, and maximum values for the year 2003.

Table A-14. Facility 6 Key Operational Parameters and Treated Effluent Water Quality.

Key Operational Parameters		
Parameter	**Unit**	**Concentration**
pH	-	Range = 7.4 – 8.0
BOD loading	kg/day	Avg = 17,327 Range = 8,656 – 28,571
BOD concentration	mg/L	Avg = 244 Range = 125 – 421
COD concentration	mg/L	NA
HRT primary	Hours	NA
Mixed liquor total suspended solids	mg/L	Avg = 1,963 Range = 1,760 – 2,173
Mixed liquor aeration time	Hours	Range = 10.5 – 11.5
SRT	Days	Range = 4 – 16
Treated Effluent Parameters		
Parameter	**Unit**	**Concentration**
pH	-	Range = 6.3 – 7.8
Temperature	°C	Range = 18.2 – 29.8
Conductivity	µS/cm	NA
BOD concentration	mg/L	Range = < 2 – 4
COD concentration	mg/L	NA
TOC/DOC	mg/L	NA
NH_4-N	mg/L	Range = < 0.10 – 3.68
NO_3^--N	mg/L	Avg = 8.2 Range = 2.48 – 14.6
PO_4^{3-}-P (reported semi-annually)	mg/L	Avg = 0.261 Range = 0.07 – 1.86

Sampling Strategy

During occurrence monitoring, two 24-hour composite samples were collected from Facility 6 to evaluate the presence and overall removal of the selected target compounds. These samples included raw sewage at the discharge of the grit basin, and tertiary effluent after disinfection. During performance monitoring, three 4-hour composite samples were collected to examine the removal efficiencies of the target compounds during individual unit operations. These samples included raw sewage, secondary effluent at the discharge of the secondary clarifier, and tertiary effluent after disinfection. The 4-hour composite samples at each unit operation generally considered the hydraulic retention time of each unit. Sampling information is summarized below in Table A-15.

Table A-15. Facility 6 Sampling Information.

Sample Location	Samples (type and date)	Samples (type and date)	Parameters	Analytic Responsibility	Container
Raw sewage after grit basin discharge	24-hr composite	4-hr composite Time: 0	pH, T, conductivity, NH_4-N, NO_3-N, PO_4-P	Facility 6	NA (on-site)
	June 2005	December 2005	pH, T, conductivity, DOC, UV	CSM	1 x 250 mL glass amber bottle
			pH, T GC-MS; LC/MS-MS	SNWA	2 x 1 L glass amber bottle
Secondary treated effluent at secondary clarifier discharge	NA	4-hr composite Time: 0 + 10 hrs	pH, T, conductivity, NH_4-N, NO_3-N, PO_4-P, MLTSS	Facility 6	NA (on-site)
		December 2005	pH, T, conductivity, DOC, UV	CSM	1 x 250 mL glass amber bottle
			pH, T GC-MS; LC/MS-MS	SNWA	2 x 1 L glass amber bottle
Tertiary effluent after disinfection	24-hr composite	4-hr composite Time: 0 + 11 hrs	pH, T, conductivity, NH_4-N, NO_3-N, PO_4-P	Facility 6	NA (on-site)
	June 2005	December 2005	pH, T, conductivity, DOC, UV	CSM	1 x 250 mL glass amber bottle
			pH, T GC-MS; LC/MS-MS	SNWA	2 x 1 L glass amber bottle

Facility 7:

Treatment

In 2003, Facility 7 had a design capacity of 91 mgd serving approximately 676,000 people, but was operated at an average capacity of 61.7 mgd. Approximately 90% of the wastewater Facility 7 receives is domestic and 10% is industrial. Stormwater sewers are not serviced by the facility. Drinking water is supplied by a combination of groundwater and surface water.

Treatment at Facility 7 consists of grit removal and flocculation/coagulation prior to primary clarification, biological nutrient removal with activated sludge treatment, secondary clarification, tertiary multi-media filtration (anthracite, sand, and gravel), and chlorine disinfection (sodium hypochlorite) with subsequent chlorine removal using sodium bisulfite. The facility employs on-site solids treatment with 30-day anaerobic digestion followed by landfill disposal. The process schematic for Facility 7 is illustrated below in Figure A-8. The hydraulic retention time during primary treatment varied between 2 and 4 hours at 61.7 mgd. The average mixed liquor aeration time also varied between 2 and 4 hours and the solids retention time varied between 10 and 15 days. Flow through the multi-media filters averaged between 53 and 72 gpm.

Water Quality

Key operational parameters and water quality of treated effluent at Facility 7 are summarized in Table A-16, which presents monthly average, minimum, and maximum values for the year 2003.

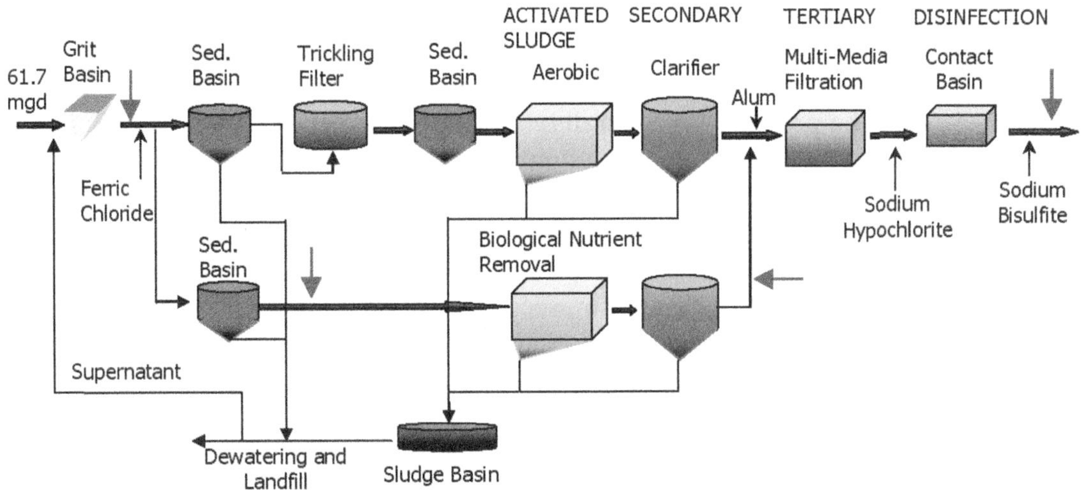

Figure A-8. Process Schematic of Facility 7.

Table A-16. Facility 7 Key Operational Parameters and Treated Effluent Water Quality.

Key Operational Parameters		
Parameter	**Unit**	**Concentration**
pII	-	NA
BOD loading	kg/day	Avg = 12,086 Range = 7,506 – 17,668
BOD concentration	mg/L	Avg = 59.3 Range = 43.3 – 75.4
COD concentration	mg/L	NA
HRT primary	Hours	Range = 2 - 4
Mixed liquor total suspended solids	mg/L	Avg = 2,610 Range = 2,142 – 3,291
Mixed liquor aeration time	Hours	Range = 2 - 4
SRT	Days	Range = 10 - 15
Treated Effluent Parameters		
Parameter	**Unit**	**Concentration**
pH	-	Range = 6.22 – 7.46
Temperature	°C	Range = 18.2 – 29.8
Conductivity	μS/cm	NA
BOD concentration	mg/L	Range = < 2 – 4
COD concentration	mg/L	NA
TOC/DOC	mg/L	NA
NH_4-N	mg/L	Range = < 0.10 – 3.68
NO_3^--N	mg/L	Avg = 8.2 Range = 2.48 – 14.6
PO_4^{3-}-P (reported semi-annually)	mg/L	Avg = 0.261 Range = 0.07 – 1.86

Sampling Strategy

During occurrence monitoring, two 24-hour composite samples were collected from Facility 7 to evaluate the presence and overall removal of the selected target compounds. These samples include raw sewage prior to coagulation addition and the primary sedimentation basin, and tertiary effluent after disinfection and chlorine removal. During Task 7, four 4-hour composite samples will be collected to examine the removal efficiencies of the target compounds during individual unit operations. These samples include raw sewage, primary effluent at the discharge of the primary sedimentation basin, secondary treated effluent at the discharge of the secondary clarifier, and tertiary effluent after chlorine removal. The 4-hour composite samples at each unit operation generally consider the hydraulic retention time of each unit. Sampling information is summarized below in Table A-17.

Table A-17. Facility 7 Task 6 and Task 7 Sampling Locations and Target Parameters.

Sample Location	Samples (type and date)	Samples (type and date)	Parameters	Analytic Responsibility	Container
Raw sewage after grit basin discharge and prior to ferric chloride addition	24-hr composite June 2005	4-hr composite Time: 0 December 2005	pH, T, conductivity, NH_4-N, NO_3-N, PO_4-P	Facility 7	NA (on-site)
			pH, T, conductivity, DOC, UV	CSM	1 x 250 mL glass amber bottle
			pH, T GC-MS; LC/MS-MS	SNWA	2 x 1 L glass amber bottle
Primary effluent at primary sedimentation basin discharge	NA	4-hr composite Time: 0 + 2 hrs December 2005	pH, T, conductivity, NH_4-N, NO_3-N, PO_4-P, MLTSS	Facility 7	NA (on-site)
			pH, T, conductivity, DOC, UV	CSM	1 x 250 mL glass amber bottle
			pH, T GC-MS; LC/MS-MS	SNWA	2 x 1 L glass amber bottle
Secondary treated effluent at secondary clarifier discharge	NA	4-hr composite Time: 0 + 4 hrs December 2005	pH, T, conductivity, NH_4-N, NO_3-N, PO_4-P, MLTSS	Facility 7	NA (on-site)
			pH, T, conductivity, DOC, UV	CSM	1 x 250 mL glass amber bottle
			pH, T GC-MS; LC/MS-MS	SNWA	2 x 1 L glass amber bottle
Tertiary effluent after disinfection and sodium bisulfite addition	24-hr composite June 2005	4-hr composite Time: 0 + 5 hrs December 2005	pH, T, conductivity, NH_4-N, NO_3-N, PO_4-P	Facility 7	NA (on-site)
			pH, T, conductivity, DOC, UV	CSM	1 x 250 mL glass amber bottle
			pH, T GC-MS; LC/MS-MS	SNWA	2 x 1 L glass amber bottle

APPENDIX B

LC/MS-MS Method Development Data

Table B-1. LC/MS-MS Compound Selection Summary.

Compound	Infusion					
	Positive	Negative	LC Method	Extraction	Conclusion	Reason for Exclusion
BHT		Yes	No		No	Very little response
Dibutyl phthalate	Yes		No		No	Blank contamination
Atrazine	Yes		Yes	Yes	Yes	N/A
Bisphenol A		Yes	Yes	Yes	Yes	N/A
Benzophenone	Yes		No		No	Blank contamination
Oxybenzone	Yes		Yes	Yes	Yes	N/A
Triclosan		Yes	Yes	Yes	Yes	N/A
t-Butyl hydroquinone		Yes	No		No	Very little response
Triclocarban		Yes	Yes	Yes	Yes	N/A
C11 LAS		Yes	No		No	Blank contamination
C12 LAS		Yes	No		No	Blank contamination
Simazine	Yes		Yes	Yes	Yes	N/A
DEET	Yes		Yes	Yes	Yes	N/A
Hydrocortisone	Yes		Yes	Yes	Yes	N/A
BHA		Yes	Yes	Yes	Yes	N/A
Vanillin		Yes	No		No	Blank contamination
o-Phenylphenol		Yes	Yes	Yes	Yes	N/A
Methylene bisphenyl diisocyante	No				No	Very little response
Saccharin		Yes	Yes	No	No	Poor recovery
FD&C Yellow No. 5		Yes	No		No	Very little response
2-Phenoxyethanol	No				No	Weak ionization
Hexabromocycledodecane	No				No	Weak ionization
t-Butyl peroxybenzoate	No				No	Weak ionization
Imidazolidinyl urea		Yes	Yes	No	No	Poor recovery
p-Aminophenol	Yes		Yes	No	No	Poor recovery
3-Indolebutyric acid		Yes	Yes	Yes	Yes	N/A
Camphor	No				No	Weak ionization
Menthol	No				No	Weak ionization
Folic acid	No				No	Weak ionization
2-Methylresorcinol		Yes	No		No	Very little response
Isobutylparaben		Yes	Yes	Yes	Yes	N/A

Table B-1. LC/MS-MS Compound Selection Summary (cont'd).

Compound	Infusion Positive	Infusion Negative	LC Method	Extraction	Conclusion	Reason for Exclusion
Farnesol	No				No	Weak ionization
Avobenzone	No				No	Weak ionization
Trichloromelamine		Yes	No		No	Very little response
Acriflavine	Yes		Yes	Yes	Yes	N/A
Rotenone		Yes	No		No	Very little response
Trifluralin	No				No	Weak ionization
MGK 11	No				No	Weak ionization
Propylparaben		Yes	Yes	Yes	Yes	N/A
Cholecalciferol	Yes		No		No	Very little response

Table B-2. Compound-Dependent Parameters – ESI Positive.

Compound	Retention Time (min)	Precursor Ion Mass	Product Ion Mass	Declustering Potential (V)	Collision Energy (eV)	Collision Cell Exit Potential (V)
Acriflavine	8.6	210.26	193.05	61	45	6
Acriflavine Confirmatory	8.6	210.26	166.15	61	41	16
Hydrocortisone	12.3	363.23	120.95	66	31	10
Hydrocortisone Confirmatory	12.3	363.23	91.15	66	77	6
Simazine	12.4	202.13	132.15	51	29	12
Simazine Confirmatory	12.4	202.13	104.05	51	35	10
Atrazine	12.8	216.11	173.76	71	27	14
Atrazine Confirmatory	12.8	216.11	103.88	71	45	10
DEET	12.7	192.17	118.93	61	25	8
DEET Confirmatory	12.7	192.17	91.02	61	45	10
Oxybenzone	13.7	229.03	150.94	46	29	16
Oxybenzone Confirmatory	13.7	229.03	104.98	46	29	10
$^{13}C_3$-Simazine	12.4	205.12	105.95	41	35	8
$^{13}C_3$-Atrazine	12.8	219.11	176.76	71	27	14

Table B-3. Compound-Dependent Parameters – ESI Negative.

Compound	Retention Time (min)	Precursor Ion Mass	Product Ion Mass	Declustering Potential (V)	Collision Energy (eV)	Collision Cell Exit Potential (V)
3-Indolebutyric acid	11.0	202.07	157.85	-45	-20	-13
3-Indolebutyric acid Confirmatory	11.0	202.07	115.95	-45	-22	-7
Bisphenol A	12.5	227.00	211.85	-60	-28	-11
Bisphenol A Confirmatory	12.5	227.00	132.85	-60	-34	-7
Propylparaben	12.7	179.01	92.05	-45	-32	-13
Propylparaben Confirmatory	12.7	179.01	135.85	-45	-22	-13
o-Phenylphenol	13.0	169.07	114.95	-60	-40	-9
o-Phenylphenol Confirmatory	13.0	169.07	141.05	-60	-34	-9
Isobutylparaben	13.0	193.16	92.05	-50	-34	-15
Isobutylparaben Confirmatory	13.0	193.16	135.85	-50	-24	-1
BHA	13.2	179.10	163.85	-30	-18	-13
BHA Confirmatory	13.2	179.10	149.05	-30	-32	-7
Triclocarban	13.8	312.97	159.75	-40	-18	-13
Triclocarban Confirmatory	13.8	312.97	125.95	-40	-30	-7
Triclosan	13.9	286.86	35.15	-45	-38	-15
Triclosan Confirmatory	13.9	286.86	241.10	-45	-10	-21
$^{13}C_{12}$-Bisphenol A	12.5	239.25	224.05	-60	-26	-17
$^{13}C_{6}$-o-Phenylphenol	13.0	175.06	121.15	-65	-42	-9
$^{13}C_{12}$-Triclosan	13.9	299.07	35.05	-40	-38	-3

Table B-4. Source-Dependent Parameters.

	ESI Positive	ESI Negative
Collision Gas (psig)	8	10
Curtain Gas (psig)	10	18
Ion Source Gas 1 – Nebulizer Gas (psig)	50	50
Ion Source Gas 2 – Turbo Gas (psig)	50	50
Ionspray Voltage (V)	5500	-4500
Temperature (°C)	600	500
Probe X-axis position (mm)	5	5
Probe Y-axis position (mm)	5	5
Entrance Potential (V)	10	10

Table B-5. Reporting Limits and Recoveries.

Compound	Reporting Limit (ng/L)	Average Recovery (%)
		n=5
ESI Positive		
Acriflavine	1.0	68
Hydrocortisone	1.0	92
Simazine	1.0	92
Atrazine	1.0	93
DEET	0.1	81
Oxybenzone	1.0	82
ESI Negative		
3-Indolebutyric acid	1.0	60
Bisphenol A	10	98
Propylparaben	0.25	116
o-Phenylphenol	10	89
Isobutylparaben	0.25	110
BHA	1.0	91
Triclocarban	0.25	77
Triclosan	1.0	92

APPENDIX C

GC/MS-MS METHOD DEVELOPMENT DATA

Table C-1. Target Compound Quantification (1) and (2) Confirmatory Transitions.

Compound	Precursor ion	Product ion
GC/MS-MS		
Camphor (1)	95	67
Camphor (2)	108	93
Menthol (1)	95	67
Menthol (2)	123	81
Phenoxyethanol (1)	138	94
Phenoxyethanol (2)	94	66
Methylresorcinol (1)	124	95
Methylresorcinol (2)	95	77
Vanillin (1)	151	65+77+108
BHA (1)	165	137
BHA (2)	180	165
BHT (1)	205	177
BHT (2)	220	205
o-Phenylphenol (1)	170	141
o-Phenylphenol (2)	141	115
Deet (1)	190	145
Deet (2)	119	91
MGK-11 (1)	175	97
MGK-11 (2)	97	79
Benzophenone (1)	105	77
Trifluralin (1)	306	264
Trifluralin (2)	264	206
Dibutyl phthalate (1)	149	121
Dibutyl phthalate (2)	223	149
Hexabromododecane (1)	239	157
Hexabromododecane (2)	319	237

Table C-2. Expected Method Reporting Limit (MRL) and Average Percent Recovery.

Compound	MRL	Average Percent Recovery DI water	Average Percent Recovery Raw water	Average Percent Recovery Wash water
GC/MS-MS	ppt	%	%	%
Camphor	50	79	76	81
Menthol	50	81	86	105
Phenoxyethanol	100	66	81	98
m-Resorcinol	100	36	62	45
Vanillin	100	59	54	109
BHA	25	64	88	127
BHT	25	79	93	111
o-Phenylphenol	25	83	95	138
DEET	25	75	90	111
MGK-11	25	76	89	126
Benzophenone	25	97	101	163
Trifluralin	25	82	97	153
Dibutyl phthalate	25	97	69	74
Hexabromododecane	100	72	95	86

Table C-3. Compounds Detected in Unspiked Samples.

Compound	Unspiked DI water	Unspiked raw water	Unspiked Wash water
GC/MS-MS	ppt	Ppt	ppt
Camphor	<50.0	<50.0	<50.0
Menthol	<50.0	<50.0	<50.0
Phenoxyethanol	<100.0	<100.0	102
m-Resorcinol	<100.0	<100.0	<100.0
Vanillin	<100.0	<100.0	82[a]
BHA	<25.0	<25.0	18[a]
BHT	<25.0	<25.0	26
o-Phenylphenol	<25.0	<25.0	<25.0
DEET	<25.0	<25.0	140
MGK-11	<25.0	<25.0	<25.0
Benzophenone	<25.0	<25.0	223
Trifluralin	<25.0	<25.0	<25.0
Dibutyl phthalate	125	239	251
Hexabromododecane	<100.0	<100.0	<100.0

[a] Value reported is below reporting limit, but shown to represent that it is detected.

Table C-4. Comparison of Compound Recoveries with Different SPE Methods.

Compound	Methylene Chloride	Methanol, MTBE, and methylene chloride	MTBE and methylene chloride
GC/MS-MS	%	%	%
Camphor	67	46	79
Menthol	70	55	81
Phenoxyethanol	49	22	66
m-Resorcinol	0	0	36
Vanillin	76	37	59
BHA	48	56	64
BHT	58	42	79
o-Phenylphenol	70	64	83
DEET	63	41	75
MGK-11	66	55	76
Benzophenone	85	78	97
Trifluralin	69	67	82
Dibutyl phthalate	47	53	97
Hexabromododecane	66	48	72

APPENDIX D

HIGH PRODUCTION CHEMICAL DATABASE

The High Production Chemical Database developed during this study provided the basis for the target compound selection and tier ranking. The database is established in the form of a Microsoft Excel spreadsheet summarizing information on:

a) Compounds used in household products
b) CAS number
c) Synonym
d) Areas of application of the chemical
e) Production volumes (according to International Council of Chemical Associations (ICCA), U.S. EPA, and EU databases)
f) Hazardous assessment (i.e., acute toxicity, chronic toxicity, teratogenicity, mutagenicity, ecotoxicity)
g) Environmental fate (e.g., biodegradation; bioconcentration constant; soil adsorption (K_{oc}); acid dissociation constant (pK_a); Henry's Law constant (H); octanol-water partitioning coefficient (log K_{ow}); water solubility (S))
h) Chemical structure
i) Additional information (e.g., appropriateness for analytical detection, etc.)

A brief description follows for each selection criteria. More details are provided in Chapter 2.0.

a) Compounds used in household products

Household chemicals were derived from a database offered by National Institutes of Health (NIH) and the National Library of Medicine. This is the most comprehensive database we identified in our review covering different household commodities. It is based on material safety data sheets (MSDSs) offered by producers operating within the United States. Approximately 720 compounds were retrieved from that list out of eight commodities: 1) auto products, 2) inside the home, 3) pesticides, 4) home maintenance, 5) personal care/use, 6) pet care, 7) arts and craft, and 8) landscape/yard. All inorganic and organic compounds were included from this database even though only organic compounds are considered in this study. Drugs and food additives were not considered.

b) CAS numbers

Chemical abstracts service (CAS) numbers are included for all chemicals. A small group of CAS numbers represents groups or mixtures of compounds. These were not further considered due to analytical constraints.

c) Synonyms

Synonyms or popular abbreviations were included for each chemical as included in the NIH database.

d) Areas of application of the chemical

Common applications of the chemicals in household commodities are summarized.

e) Production volumes

Three databases were used to evaluate production volumes and consumption patterns:

1) ICCA (International Council of Chemical Associations): chemical is considered as HPV or otherwise of interest in two or more regions of North America, Europe, or Japan;

2) U.S. EPA Inventory Update Rule: production volume levels of 10,000 pounds or more for produced or imported chemicals in U.S. HPV chemicals in the U.S. is by definition produced or imported in amounts larger than 1 million pounds per year.

3) European Chemical Substances Information System database (ESIS): chemical being produced or imported in quantity of at least 1,000 metric tons (equivalent of 2.2 million pounds) per year in EU.

Compounds identified as HPV chemicals in the Excel database are those that are included in at least one of the three HPV chemical listings.

f) Hazardous assessment

Toxicity information for the chemicals derived from U.S. EPA's Chemical Hazard Data Availability Study (http://www.epa.gov./chemrtk/hazchem.htm). Compounds that had a positive finding for one of the six Screening Information Data Sets (SIDS) toxicity endpoints (acute toxicity, chronic toxicity, teratogenicity, mutagenicity, ecotoxicity, and environmental fate) were indicated with a "1", compounds that had no findings were indicated with "0".

g) Environmental fate

This section summarizes information on biodegradation, bioconcentration, soil adsorption, acid dissociation constants, Henry's Law constants, octanol-water partitioning coefficients, and water solubility, as available. Information derives mainly from two data sources, the Syracuse Research Corporation (2004) and U.S. EPA's software program EpiSuite (2000).

h) Chemical structure

Chemical structures are provided as available.

i) Comments

Here relevant comments are included for specific compounds such as choice of analytical method, appropriateness for analysis, or compound classification (solvents, semi-volatiles, inorganic compounds, etc.). Color codes were assigned to the compounds to make the selection approach clear:

Light yellow:	HPV chemical
Purple:	inorganic chemical
Blue:	solvents
Orange:	compound mixtures

APPENDIX E

QSAR/QSPR MODELS

Table E-1. Programs for Estimating Descriptor Parameters.

Program	Manufacturer	Input	Output	Operating System
Cranium	Molecular Knowledge Systems, Inc.	structure	20 physical property descriptors	Windows
Hyperchem	Hypercube, Inc.	structure	quantum mechanical descriptors	Windows
CLIP	Institute of Medicinal Chemistry of the University of Lausanne	structure	hydrophobicity and molecular lipophilicity descriptors	
ChemDB	TimTec, Inc.	structure	pKa and hydrophobicity and H-bond descriptors	Windows
IA_LogP	Interactive Analysis	structure	LogKow	Windows
GRID	Molecular Discovery	structure	energetically favorable binding site descriptors	Unix
mitools	Molinspiration	structure	LogKow, polar surface area, hydrogen bonding descriptors	Windows Unix (online version; free 100 cmpd/ mo)
PhysChem	ACD	structure	LogKow, Log D, pK_a	Windows Unix
BioLoom ClogP	Biobyte	structure	hydrophobic and molecular refractivity descriptors	Windows
Spartan	Wavefunction	structure	quantum mechanical descriptors	Mac

Table E-1. Programs for Estimating Descriptor Parameters, Continued.

Program	Manufacturer	Input	Output	Operating System
SPARC (public domain)	Carreira, L.A. University of Georgia	structure	pKa, Henry's law constant, polarizability	Online (free)
Dragon	Talete	structure	1,600 molecular descriptors	Windows Unix
Gaussian	Gaussian, Inc.	structure	quantum mechanical descriptors	Unix
ADF	Scientific Computing and Modeling	structure	quantum mechanical descriptors	Windows Mac Unix
Graph III		structure	molecular connectivity indices	
UNIFAC (modified)	Rogers (1994) (dissertation)	structure	Henry's law constant	Windows
ChemID*plus*	National Library of Medicine; Specialized Information Services	structure	pK_a, log K_{ow}, log D	Online
EPI Suite (public domain)	U.S. EPA Syracuse Research Corporation	structure	Henry's law constant, log K_{ow}, aerobic biodegradability, soil sorption, hydrolysis, removal by wastewater treatment	Windows

Table E-2. Programs for Estimating Descriptor Parameters and Developing QSAR/QSPR Models.

Program	Manufacturer	Input	Output	Operating System
Chem 3D CS MOPAC; ChemSAR	Adept Scientific	structure and endpoint data	quantum mechanical descriptors and QSAR development	Windows
Cerius^2 Quantum Mechanics Workbench and C^2.QSAR+	Accelrys Inc.	Structure and endpoint data	molecular descriptors and QSAR development	Unix
QSARIS	SciVision	Structure and endpoint data	molecular descriptors and QSAR development	Windows
SYBYL	Tripos	Structure and endpoint data	dependent on modules – very wide ranging and potential for QSAR development	Unix
GOLPE	MIA GOLPE can be distributed only to TRIPOS customers with a valid license for SYBYL	descriptors and endpoint data	QSAR development	Unix
Topkat	Accelrys Inc.	Structure and endpoint data	predict aerobic biodegradation and QSAR development	Windows
CODESSA PRO AMPAC	Semichem, Inc.	Structure and endpoint data	quantum mechanical descriptors and QSAR development	Windows Mac Unix
QSAR CDK, JOELib, R (public domain; unreleased)	SOURCEFORGE	Structure and endpoint data	molecular descriptors; QSAR development	Windows Unix
QuaSAR	Chemical Computing Group Inc.	Structure and endpoint data	quantum mechanical descriptors and QSAR development	Windows Mac Unix
Schrödinger QikProp, Jaguar, Strike	Schrödinger, Inc.	Structure and endpoint data	pK_a, quantum mechanical descriptors and QSAR development	Unix

REFERENCES

Alikhanidi, S. and Takahashi, Y. (2004) Pesticide persistence in the environment – collected data and structure-based analysis, *J. Comput. Chem., Japan.* 3:2:59-70.

American Water Works Association, American Public Health Association, Water Environment Federation (1998) *Standard Methods for the Examination of Water and Wastewater* 20th edition, American Public Health Association, Washington, D.C.

Atkinson, R. (1986) Kinetics and mechanisms of the gas-phase reactions of the hydroxyl radical with organic compounds under atmospheric conditions. *Chem. Rev.* 86:1:69-201.

Atkinson, R. (1987) A structure-activity relationship for the estimation of rate constants for the gas-phase reactions of OH radicals with organic compounds. *Int. J. Chem. Kinet.* 19:799-828.

Atkinson, R. (1988) Estimation of gas-phase hydroxyl radical rate constants for organic chemicals. *Environ. Toxicol. Chem.* 7:435-442.

Banerjee, S., Howard, P.H., Rosenberg, A.M., Dombrowski, A.E., Sikka, H., and Tullis, D.L. (1984) Development of a general kinetic model for biodegradation and its application to chlorophenols and related compounds. *Environ. Sci. Technol.*18:416-422.

Bartolotti, L.J. and Edney, E.O. (1994) Investigation of the correlation between the energy of the highest occupied molecular orbital (HOMO) and the logarithm of the OH rate constant of hydrofluorocarbons and hydrofluoroethers, *International Journal of Chemical Kinetics.* 26:9:913-920.

Bellona, C. (2007) Employing nanofiltration and ultra low-pressure reverse osmosis for drinking water augmentation: Efficiency of operation and removal of emerging organic contaminants. Doctoral Disseration, Environmental Science and Engineering. Colorado School of Mines.

Blanco, M. a. B. and Teruel, M.A. (2007) Rate constants of the reactions of CF2ClC(O)OCH3 and CF2ClC(O)OCH2CH3 with OH radicals and Cl atoms at atmospheric pressure, *Chemical Physics Letters,* 441:1-3:1-6.

Blum, D.J.W., Suffet, I.H., and Duguet, J.P. (1994) Quantitative structure-activity relationship using molecular connectivity for the activated carbon adsorption of organic chemicals in water, *Wat. Res.* 28:3:687-699.

Boethling, R.S. (1986) Application of molecular topology to quantitative structure-biodegradability relationships, *Environ. Toxicol. Chem.* 5:797-806.

Boethling, R.S., Howard, P.H., Meylan, W., Stiterler, H., Beauman, H., and Tirado, N. (1994) Group contribution methods for predicting probability and rate of aerobic biodegradation, *Environ. Sci. Technol.* 28:3:459-465.

Brasquet, C., Subrenat, E., and Le Cloirec, P. (1997) Selective adsorption on fibrous activated carbon of organics from aqueous solution: correlation between adsorption and molecular structure, *Wat. Sci. & Technol.* 35:7:251-259.

Brasquet, C. and Le Cloirec, P. (1999) QSAR for organics adsorption onto activated carbon in water: what about the use of neural networks, *Wat. Res.* 33:17:3603-3608.

Business Service Center (2003): Banned Consumer Chemicals, http://www.cbsc.org/alberta/search/display.cfm?Code=2266&coll=FE_FEDSBIS_E (Last updated 9-19-2003).

Chemicals Evaluation and Research Institute (CERI): Biodegradation and bioaccumulation data of existing chemicals, http://qsar.cerij.or.jp/cgi-bin/DEGACC/index.cgi?e.

Clark, B., Henry, J.G., and MacKay, D. (1995) Fugacity Analysis and Model of Organic Chemical Fate in a Sewage Treatment Plant, *Environ. Sci. Technol.* 29(5):1488-1494.

Cooper, D.L. and Cunningham, T.P. (1992) Tropospheric lifetimes of potential CFC replacements: rate coefficients for reaction with the hydroxyl radical, *Atmospheric Environment.* 26A:7:1331-1334.

Crittenden, J., Sanongraj, S., Bulloch, J., Hand, D., Rogers, T., Speth, T. and Ulmer, M. (1999) Correlation of aqueous-phase adsorption isotherms. Environ. Sci . Technol. 33:17:2926-2933.

Dearden, J.C. and Nicholson, R.M. (1986) The prediction of biodegradability by the use of quantitative-structure-activity relationships: correlation of biological oxygen demand with atomic charge difference, *Pest. Sci.* 17:305-310.

Dearden, J.C. and Nicholson, R.M. (1987) Correlation of biodegradability with atomic charge difference and superdelocalizability, In *QSAR in Environmental Toxicology – II,* Kaiser, K.L.E., Ed., Reidel, Dordrecht, pp. 83-89.

Dearden, J.C. and Cornin, M.T.D. (**1996**) Quantitative structure-biodegradability studies: and investigation of the MITI aromatic compound data-base, In *Biodegradability Prediction,* Peijnenburg, W.J.G.M. and Damborsky, J., Eds., Kluwer Academic Publishers: Boston, MA, pp 93-104.

Dobbs, R.A., Wang, L., and Govind, R. (**1989**) Sorption of toxic organic compounds on wastewater solids: correlation with fundamental properties, *Environ. Sci. Technol.* 23:9:1092-1097.

Dominiak, M.F. (2004), U.S. Environmental Protection Agency, personal information.

Environment Canada (2003a): The Substances List. Search Engine for Chemicals and Polymers, http://www.ec.gc.ca/substances/nsb/eng/cas_e.htm, (last updated: 3-27-2003).

Environment Canada (2003b): Existing Substances Evaluation, http://www.ec.gc.ca/substances/ese/eng/esehome.cfm, (last updated: 5-13-2003)

Environment Canada (2004): Published Orders. Significant New Activity (SNAs), http://www.ec.gc.ca/substances/nsb/eng/snac_e.htm (last updated: 4-13-2004)

Environmental Impact Initiative (2003): http://www.hc-sc.gc.ca/ear-ree/index_e.html, (last updated: 11-10-2003)

European Chemicals Bureau (ECB) (2004): European Chemicals Substances Information System (ESIS), http://ecb.jrc.it

European Commission (1999): Public availability on data of EU high production volume chemicals, Joint Research Centre, Institute for Health and Consumer Protection European Chemicals Bureau, Allanou, R., Hansen, B.G., van der Bilt, Y.

European Commission (2004): Chemicals. (last updated: 8-23-2004) http://europa.eu.int/comm/environment/chemicals

European Commission (2007): Regulation (EC) No 1907/2006 of the European Parliament and of the Council of 18 December 2006 concerning the Registration, Evaluation, Authorisation and Restriction of Chemicals (REACH). http://eur-lex.europa.eu/LexUriServ/LexUriServ.do?uri=CELEX:32006R1907:EN:NOT

Food and Drug Administration (FDA) (2004a): *Electronic Orange Book. Approved Drug Products with Therapeutic Equivalence Evaluations,* http://www.fda.gov/cder/ob/default.htm, (last updated: 2004)

Food and Drug Administration (FDA) (2004b): Center for Drug Evaluation and Research, Drug Information, http://www.fda.gov/cder/drug/default.htm, last updated: November 24, 2004.

Food and Drug Administration (FDA): Title 21 Food and Drugs, Chapter 1: Food and Drug Administration, Department of Health and Human Services, http://www.access.gpo.gov/nara/cfr/waisidx_01/21cfrv3_01.html.

Gawlik, B.M., Sotiriou, N. Feicht, E.A., Schulte-Hostede, S., and Kettrup, A. (1997) Alternatives for the determination of the soil adsorption coefficient, KOC, of non-ionic organic compounds – a review, *Chemosphere*, 34:12:2525-2551.

Gould, J.P. (1987) Correlations between chemical structure and ozonation kinetics: preliminary observations. *Ozone Science and Engineering.* 9:207-216.

Greaves, A.J., Churchley, J.H., Hutchings, M.G., Phillips, D.A.S., and Taylor, J.A. (2001) A chemometric approach to understanding the bioelimination of anionic, water-soluble dyes by a biomass using empirical and semi-empirical molecular descriptors, *Wat. Res.* 35:5:1225-1239.

Haag, W.R. and Yao, D. (1992) Rate constants for reaction of hydroxyl radicals with several drinking water contaminants. Environmental Science and Engineering 26, 5, 1005-1013.

Heidler, J., Sapkota, A., and Halden, R. (2006) Partitioning, Persistence, and Accumulation in Digested Sludge of the Topical Antiseptic Triclocarban during Wastewater Treatment, *Environ. Sci. Technol., 40*(11):3634 - 3639;

Hiromatsu, K., Yakabe, Y., Katagiri, K., and Nishihara, T. (2000) Prediction for biodegradability of chemicals by an empirical flowchart, *Chemosphere*, 41:1749-1754.

Hoigne, J. and Bader, H. (1983) Rate constants of reactions of ozone with organic and inorganic compounds in water-II, *Wat. Res.* 17:185-194.

Howard, P.H., Boethling, R.S., Stitler, W.M., Leylan, W.M., Hueber, A.E., Beauman, J.A., and Larosche, M.E. (1992) *Environ. Toxicol. Chem.* 11:593-603.

Hsieh, H-N. and Mukherjee, S. (2003) A QSAR model for desorption of halogenated aliphatics from biosolids, *Advances in Environmental Research.* 7:511-520.

Hu, J-Y., Morita, T., Magara, Y., and Aizawa, T. (2000) Evaluation of reactivity of pesticides with ozone in water using the energies of frontier molecular orbitals, *Wat. Res.* 34:8:2215-222.

International Council of Chemical Associations (ICCA) (2004): ICCA HPV working list, http://www.cefic.org/activities/hse/mgt/hpv/hpvinit.htm (last updated June 2004)

Kahn, I., Fara, D., Karelson, M., and Maran, U. (2005) QSPR treatment of the soil sorption coefficients of organic pollutants, *J. Chem. Inf. Model.* 45:1:94-105.

Kinney, C., Furlong, E., Zaugg, S., Burkhardt, M., Werner, S., Cahill, J., and Jorgensen, G. 2006. Survey of organic wastewater contaminants in biosolids destined for land application, *Environ. Sci. Technol.* 40(23):7207-7215.

Klopman, G., Balthasar, D.M., and Rosenkranz, H.S. (1993) Application of the computer-automated structure evaluation (CASE) program to the study of structure-biodegradation relationships of miscellaneous chemicals, *Environ. Toxicol. Chem..* 12:231-240.

Kwok, E.S.C. and Atkinson, R. (1995) Estimation of hydroxyl radical reaction rate constants for gas-phase organic compounds using a structure-reactivity relationship: An update. *Atmospheric Environment,* 29:14:1685-1695.

Landrigan, P.J. (2000): Disease of the environmental origin in American children: prospects for research and prevention. Testimony before the Committee on Appropriations U.S. House of Representatives. Washington, D.C.

Lei, H. and Snyder, S.A. (2007) 3D QSPR models for the removal of trace organic contaminants by ozone and free chlorine. Water Research 41, 18, 4051-4060.

Li, L., Quinlivan, P.A., Knappe, D.R.U. (2005) Predicting adsorption isotherms for aqueous organic micropollutants from activated carbon and pollutant properties. *Environ. Sci. Technol.* 39:9:3393-3400.

Loonen, H., Lindgren, F., Hansen, B., Karcher, W., Niemelä, J., Hiromatsu, K., Takatsuki, M., Peijnenburg, W., Roriji, E., and Struijś, J. (1999) Prediction of biodegradability from chemical structure: modeling of ready biodegradation test data, *Environ. Toxicol. Chem.* 18:8:1763-1768.

Luehrs, D.C., Hickey, J.P., Nilsen, P.E., Godbole, K.A., and Rogers, T.N. (1996) Linear solvation energy relationship of the limiting partition coefficient of organic solutes between water and activated carbon. *Environ. Sci. Technol.* 30:1:143-152.

Magnuson and Speth (2005) Quantitative structure-property relationships for enhancing predictions of synthetic organic chemical removal from drinking water by granular activated carbon. *Environ. Sci . Technol.* 39:19:7706-7711.

McElroy, J. (2005) Adsorption of substituted aromatic compounds by activated carbon: a mechanistic approach to quantitative structure activity relationships. Thesis, University of Florida.

Ministry of the Environment (2004): http://www.env.go.jp/en.

National Institute for Public Health and the Environment (RIVM) (1997): Prediction of environmental degradation rates for High Production Volume Chemicals (HPVC) using Quantitative Structure-Activity Relationships, http://www.rivm.nl/bibliotheek/rapporten/719101030.html.

National Institutes of Health (NIH) (2004): http://hpd.nlm.nih.gov, (last updated 5-12-2004)

National Institutes of Health (NIH) & U.S. National Library of Medicine (2004): Medline Plus. Drug information, http://www.nlm.nih.gov/medlineplus/druginformation.html, last updated August 2004.

Niemi, G.J., Veith, G.D., Regal, R.R., and Vaishnav, D.D. (1987) Structural features associated with degradable and persistent chemicals, *Environ. Toxicol. Chem.* 6:515-527.

Nirmalakhandan, N.N. and Speece, R.E. (1988) QSAR model for predicting Henry's constant, *Environ. Sci. Technol.* 22:11:1349-1357.

Okey, R.W. and Stensel, H.D. (1996) A QSAR-based biodegradability model-a QSBR, *Wat. Res.* 30:9:2206-2214.

Organization of Economic Cooperation and Development (OECD) (2001): Environmental Outlook for the Chemicals Industry. Paris.

Paris, D.F., Wolfe, N.L. and Steen, W.C. (1982) *Applied Environmental Microbiology*. 44:153-158.

Parsons, J.R., Commandeur, L.C.M., van Eyseren, H.E., and Govers, H.A.J. (1991) QSARs and PARs for biodegradation of PCBs, *Sci. for the Total Environment*. 109-110:275-281.

Peijnenburg, W. (1991) The use of quantitative structure-activity relationships for predicting rates of environmental hydrolysis processes, *Pure & Appl. Chem.* 63:11:1667-1676.

Peijnenburg, W. (1994) Structure-activity relationships for biodegradation: a critical review, *Pure & Appl. Chem.* 66:9:1931-1941.

Percival, C.J., Marston, G., and Wayne, R.P. (1995) Correlations between rate parameters and calculated molecular properties in the reactions of the hydroxyl radical with hydrofluorocarbons, *Atmospheric Environment* 29:3:305-311.

Procter & Gamble (2004): Ingredient safety information, http://www.scienceinthebox.com/en_UK/glossary/perfume_en.html.

Product Safety Program (PSP) (2004): Healthy Environments and Consumer Safety Branch (HECS), www.hc-sc.gc.ca/hecs-sesc/psp/index.htm, (last updated 6-29-2004)

Raymond, J.W., Rogers, T.N., Shonnard, D.R., and Kline, A.A. (2001) A review of structure-based biodegradation estimation methods, *J. of Hazard. Mater.* B84:189-215.

Rodriguez, G., Buonora, S., Knoell, T., Phipps, Jr. D., and Ridgway, H. (2004) *Rejection of Pharmaceuticals by Reverse Osmosis Membranes: Quantitative Structure Activity Relationship (QSAR) Analysis*, Final Report, NWRI Project 01-EC-002.

Rorije, E., Germa, F., Philipp, B., Schink, B., and Beimborn, D.B. (2002) Prediction of biodegradability from structure: imidazoles, *SAR and QSAR in Environmental Research*. 13:1:199-204.

Rx.com Partners (2004): Drug search, http://www.rx.com/rx.asp?n=0&i=zx, (last updated 2004)

Sabljić, A. (1987) On the prediction of soil sorption coefficients of organic pollutants from molecular structure: application of molecular topology model, *Environ. Sci. Technol.* 21:4:358-366.

Sabljić, A. and Güsten, H. (1989) Predicting Henry's law constants for polychlorinated biphenyls, *Chemosphere*, 19:10-11:1503-1511.

Sabljić, A. and Harvatić, D. (1993) Graph III: A computer program for calculating molecular connectivity indices on microcomputers, *J. Chem. Inf. Comput. Sci.*, 33:3:292-295.

Sabljić, A. (1995) QSAR modelling of soil sorption improvements. Improvements and systematics of log K_{OC} vs. LogKow correlations, *Chemosphere*, 31:11-12:4489-4514.

Sabljić, A. (2001) QSAR models for estimating properties of persistent organic pollutants required in evaluation of their environmental fate and risk, *Chemosphere*, 43:363-375.

Sabljić, A. and Peijnenburg, W. (2001) Modeling lifetime and degradability of organic compounds in air, soil, and water systems, *Pure Appl. Chem.* 73:8:1331-1348.

Salvito et al. (2002): A framework for prioritising fragrance materials for aquatic risk assessment. *Environ. Tox. Chem.* 21(6): 1301-1308.

Seattle Daily Journal of Commerce (1997): This hazardous waste program hits home, by D. Miscolta and J. Howell. August 21, 1997, Seattle, WA.

Stanford Research Institute (2004): Chemical Economics Handbook, http://ceh.sric.sri.com.

Syracuse Research Corporation (SRC): Business Areas. Environmental Fate Data Base (EFDB), http://www.syrres.com/esc/efdb.htm, last updated: September 3, 2004.

Tang, W.Z. (2004) *Physicochemical Treatment of Hazardous Wastes*, Lewis Publishers: Washington, D.C.

Tratnyek, P.G. and Hoigne, J. (1994) Kinetics of reactions of chlorine dioxide in water-II. Quantitative structure-activity relationships for phenolic compounds, *Wat. Res.* 28:57.

United Nations Environmental Program (UNEP) (1996): Screening Information Data Set (SIDS) for High Production Volume Chemicals. (Volumes 1-4. United Nations, New York and Geneva. March 1996.

United States Environmental Protection Agency (U.S. EPA) (1998): Chemical Hazard Data Availability Study. What do we really know about the safety of high production volume chemical? EPA's Office of Pollution Prevention and Toxics. Washington, D.C.

United States Environmental Protection Agency (U.S. EPA) (2000): Exposure assessment tools and models, http://www.epa.gov/opptintr/exposure/docs/episuitedl.htm, last updated: November 5, 2004.

United States Environmental Protection Agency (U.S. EPA) (2002): Chemical hazard data availability study. (last updated 9-17-2002). www.epa.gov/chemrtk/hazchem.htm

United States Environmental Protection Agency (U.S. EPA) (2003): Inventory Update Rule. (last updated: 12-10-2003). http://www.epa.gov/opptintr/iur

United States Environmental Protection Agency (U.S. EPA) (2004a): Substance Registry System (SRS), http://www.epa.gov/srs, (last updated 3-26-2004).

United States Environmental Protection Agency (U.S. EPA) (2004b): Toxic Release Inventory (TRI) Program, http://www.epa.gov/tri/tridata/tri02/data/index.htm, last updated: June 24th, 2004.

Wargo, J. and Wargo, L.E. (2002): The State of Children's Health and Environment. Children's Health Environmental Coalition. Princeton, NJ.

Westerhoff, P., Yoon, Y., Snyder, S., and Wert, E. 2005. Fate of endocrine-disruptor, pharmaceutical, and personal care product chemicals during simulated drinking water treatment processes. *Environmental Science and Technology,* 39:6649-6663.

IWA Publishing's authorised EU representative for General Product Safety Regulations is Diane D'Arras, 15 rue Duret, 75116 Paris, France, e-mail: safety@iwap.co.uk.

Printed and bound by CPI Group (UK) Ltd, Croydon, CR0 4YY
12/05/2026
02108892-0012